PLAIN FIGURES

Cabinet Office (Management and Personnel Office)
Civil Service College

PLAIN FIGURES

MYRA CHAPMAN
in collaboration with
BASIL MAHON

LONDON · HER MAJESTY'S STATIONERY OFFICE

© *Crown copyright 1986*
First published 1986
ISBN 0 11 430001 1

Contents

	Page
Acknowledgements	7
Introduction	9
Aim of this book	9
Who is it for?	9
Why is it needed?	9
Plan of the book	9
Chapter 1: General principles for presenting data	11
Reasons for presenting data	11
Selection of medium for presenting statistics	11
Tables and charts for demonstraton. Does it work?	12
General principles for demonstration tables and charts	15
Reference tables. Are they easy to use?	16
Accuracy	16
Summary of general principles	16
Chapter 2: Summary of recommendations	17
Overview	17
Choice of medium for presentation	17
Demonstration or reference	17
For demonstration, charts or tables? Write the verbal summary first	17
Tables for specific amounts	17
Charts for comparisons	17
Which numbers to present	18
Guidelines for demonstration tables	18
Guidelines for reference tables	18
When to use charts. Demonstration only, never for reference	19
What sort of chart?	19
The role of words in communicating numbers	20
Chapter 3: Structure and style of tables	21
Introduction	21
Structure	21
Applying the rules on structure	24
Style	26
Chapter 4: Demonstration tables	35
Which numbers to present	35
Some derived statistics	35
Presentation of demonstration tables	38
Seven basic rules	39
Chapter 5: Reference tables	48
Introduction	48
Choice of categories to be tabulated	48
Which factors to put in rows and which in columns	48
Order of tabulation of categories	50
Layout of large tables	50
Errors: indication of size of intrinsic error	52
Footnotes and explanatory notes covering a set of tables	52
Definitions	54
Treatment of years	54
Key to abbreviations and special symbols used	54
Notes on method of data collection	54
Commentary on reference tables	56

	Page
Some further examples	56
Chapter 6: Charts	61
Do charts work?	61
When to use a chart	64
What sort of chart?	65
How to draw charts. General guidelines	65
Chatty charts	67
Special charts	68
Chapter 7: Effective charts for general use	69
Pie charts	69
Bar charts	72
Line graphs	84
Scatter charts	88
Isotypes	89
Chapter 8: Words	92
Don't use words alone	92
Vital role of words	92
Write clearly and concisely	92
Different sorts of commentaries and different sorts of readers	93
Verbal summary of a data display	94
Commentary on reference tables	94
Report on a statistical investigation	96
Example of a good commentary on statistical tables	98
Appendix A: Tables and charts as visual aids	104
Introduction	104
Viewgraphs	104
35mm slides	105
Wall charts	106
Appendix B: Further reading	107
Demonstration tables	107
Reference tables	107
Charts	107
Visual aids	107
Words	107
Appendix C: Research evidence	108
Summary of research evidence	108
Appendix D: Bibliography	109

Acknowledgements

Our thanks are due to all the friends and colleagues in the Government Statistical Service who commented on early drafts of *Plain Figures*, who provided examples of good and bad practice in presenting numbers and who gave us copies of departmental codes of practice in drawing up statistical tables. We are particularly grateful to the editors of *Social Trends* (HMSO) and the *Employment Gazette* (HMSO) for permission to use tables and charts from these sources. Our task would have been infinitely harder if we had been unable to use *Social Trends* as a rich store of good (and not so good) examples of statistical tables and charts.

We are particularly grateful to Richard Eason who has read the entire text—some of it several times—and suggested numerous improvements in structure, style and clarity; also to John Merchant who originally suggested that we should write the book and commented on early chapters. Finally we must offer particular thanks to Janet Outred who met our early chapters as her first real job on a word processor and dealt cheerfully with countless amendments.

A humble tribute is also due to Sir Ernest Gowers, whose immensely valuable *Complete Plain Words* (a new edition of which has recently been published) was a powerful stimulus and to Sir Bruce Fraser who so ably revised and extended the original book. We hope that this book will advance their cause a little further by encouraging a similarly disciplined, thoughtful yet simple approach to the representation of numerical information.

Introduction

Aim of this book

Most people are not statisticians. But most people are regularly confronted with figures: in newspaper articles on unemployment or inflation, in reports on the performance of companies or products, in advertisements, in trade union claims for equity. Wherever the reader is invited to make a quantitative comparison the writer thinks (correctly) that the comparison will be clearer if figures are quoted.

But clarity is not always achieved, and phrases like 'Lies, damned lies, and statistics', 'Of course you can prove anything with statistics', and 'I've never had a head for figures' have become common defensive reactions. This is a great pity. Properly used, numbers provide by far the most effective way of describing changes and making comparisons in all the above areas and in many others.

The aim of this book is to demonstrate and discuss ways of presenting numbers effectively so that their value can be realised in full. A subsidiary aim is to help the reader interpret all data more competently and confidently.

It is important to state immediately that an honest intention is assumed. This book has no handy hints for those who wish to disguise the shortcomings of their data or who wish to twist their data to support a pre-conceived argument.

Who is it for?

The book is intended for anyone who needs to present statistical information, including:

—statisticians

—economists

—scientists and specialists in other numerate disciplines

—administrators, managers and others who sometimes need to present statistical information in the course of their work.

It is written particularly for those who wish to communicate a quantitative message to a non-specialist audience.

No statistical techniques are discussed. The book deals with ways of communicating numbers effectively without using terms like 'standard deviation' and 'level of significance'. Statisticians, economists and scientists will often use formal statistical techniques to analyse data but, once they have identified the main patterns in the data, the guidelines in this book may help them to present their findings to a non-technical audience.

Although as a Civil Service College publication the book is designed primarily to meet the needs of civil servants, it naturally covers the needs of others also, since the principles of good statistical presentation hold for all applications.

Why is it needed?

Statistical presentation is fundamentally important; if it is not done properly all prior work on data collection and analysis is likely to be wasted. Despite this, presentation is seldom taught in formal academic courses and has few generally known principles or standards. Consequently it is frequently done badly through lack of method, discipline, or thought about its purpose.

Badly presented statistical information can mislead people and may lead to mistakes in further calculations; it may waste people's time, and no matter how relevant, is likely to be ignored. Conversely, well presented information can be assimilated quickly and accurately and can therefore be used with ease and confidence.

For these reasons a book is needed which will encourage people to think more about their objectives in presenting quantitative information, and which will provide useful principles and guidelines on effective methods of presentation.

This book brings together advice and research findings on statistical presentation which have appeared in a variety of sources: from articles published in the 1920s to recent work on tables and graphs by A S C Ehrenberg.

Plan of the book

The book starts with a chapter on the general principles of statistical presentation, differentiating clearly between data presented for reference purposes and data presented to illustrate an argument.

The second chapter contains a summary of the arguments and recommendations developed throughout the book. This chapter is included to provide a summary of the principles of good data presentation for those with limited time and to direct such readers to chapters which may be of immediate personal interest. It also provides those who have read the whole book with a quick reminder of the important guidelines developed in later chapters.

The next three chapters are devoted to tables: first a chapter on the structure common to all statistical tables, then a chapter on tables designed to communicate specific messages and thirdly a chapter on tables designed for reference purposes.

Chapters 6 and 7 are about the use of pictures, or charts, to illustrate quantitative messages, Chapter 6 dealing with general principles and Chapter 7 discussing some specific kinds of statistical chart which can be used effectively in presenting quantitative results to a non-specialist audience. Finally the role of narrative in communicating statistical information is discussed in Chapter 8.

The book contains a great many tables and charts, some demonstrating good practice, some exhibiting various faults. For ease of reference, good tables and charts have been identified by the small symbol ☑ at the top righthand corner.

The research findings which support the guidelines offered for effective presentation of data are reviewed in Appendix C. This appendix is included to assure the reader that the guidelines discussed are not just based on personal opinions which may safely be disregarded. The guidelines stem from a variety of sources: experimental evidence, the considered opinions of experienced practitioners, evidence from the field of cognitive psychology—and, occasionally, personal opinion. Since this book offers considerably more positive advice than usual on which medium to use for which message, care has been taken to explain the basis for its recommendations.

It will be obvious to any reader who knows the work of Professor A S C Ehrenberg how much this book owes to his work on tables and graphs.

Chapter 1: General principles for presenting data

1.1 Reasons for presenting data

A set of data may appear in a report or publication for two quite distinct reasons. It may be presented for future reference, that is, to provide a range of numerical information from which other people will select the data they need for specific analyses. Or it may be presented to demonstrate a particular fact or to support an argument.

The principles to be applied in presenting data for future reference are different from those to be applied when presenting data for demonstration purposes. This is because reference data and demonstration data are produced at different stages in the process of collecting and using statistical data and are produced for different reasons. The collection and use of statistical data can be divided into the following stages:

a. collect and check the basic data

b. collate and store the data in a convenient form for later use

c. select and analyse an appropriate subset of the data

d. report findings

e. influence decisions.

The final stage is of course the ultimate purpose of collecting the data: if the data cannot be used to influence decisions the cost and effort of earlier stages are wasted.

Reference data are produced at stage b. and may be stored in computer files or recorded in tables, in the expectation that they will be used in later analyses. Many government statistical publications are primarily data storage devices which are produced at this stage. Demonstration data, however, are produced at stage d., generally in the expectation that they will be used at stage e. Different criteria must therefore be applied to the presentation of each sort of data: data designed for future reference must be presented for ease of use; data intended to support an argument or demonstrate a particular fact must be presented for ease of comprehension.

1.2 Selection of medium for presenting statistics

There are only three ways of presenting statistical information: it can appear in a table, in the form of a picture (a chart, a graph or a diagram of some sort) or in a paragraph of prose. These three media—tables, pictures and words—may be used in the following ways:

	Reference	Demonstration
Tables	Yes	Yes
Pictures	No	Yes
Words	No	Yes

If figures are being presented for *reference* purposes there is usually no sensible choice but tables. Reading numbers from a graph or chart tends to be slow and inaccurate, and words alone would be an exceedingly inefficient means of recording reference data.

For *demonstration* purposes the most effective combination of tables, pictures and words should be chosen. And here it is important to use the media in mutually supporting ways. Tables and charts very seldom 'speak for themselves' so if either is to be effective in demonstrating a particular point it should be accompanied by a verbal summary. Both tables and pictures should be clear and simple and should be included in the main text rather than in an annex at the back.

In general, when choosing how to present demonstration data, tables are best for conveying numerical values, pictures are best for conveying qualitative relationships and words are best for conveying implications for action. For example, Table 1.1 and Figure 1.1 both communicate the following messages: first that between 1978/79 and 1979/80 there was an increase in the variety of schemes used on the Youth Opportunities Programme (YOP) in Wales, although in both years Work Experience on Employers' Premises (WEEP) remained the major component: and secondly that the total number of entrants to the programme increased by almost 50 per cent (from 15,000 to 22,000 in round numbers).

Table 1.1 Entrants to Youth Opportunities Programme in Wales: by type of scheme

Wales 1978 to 1980   *Percentages*

	1978/79	1979/80
WEEP[1]	89	73
Short Training Courses	10	10
Community service		9
Project based work experience		7
Training workshops	1	1
Induction and other[2]		1
Total (100%)	15,000	22,000

[1] Work Experience on Employers' Premises
[2] Employment induction courses and other remedial and preparatory courses

The table is more likely to communicate the actual numbers effectively: the totals of 15,000 in 1978/79 and 22,000 in 1979/80, and the fact that in 1978/79, 99 per cent of all entrants joined a WEEP scheme or a Short Training Course, whereas in 1979/80 these two schemes accounted for only 83 per cent of all entrants, the

Figure 1.1   Entrants to Youth Opportunities Programme in Wales: by type of scheme

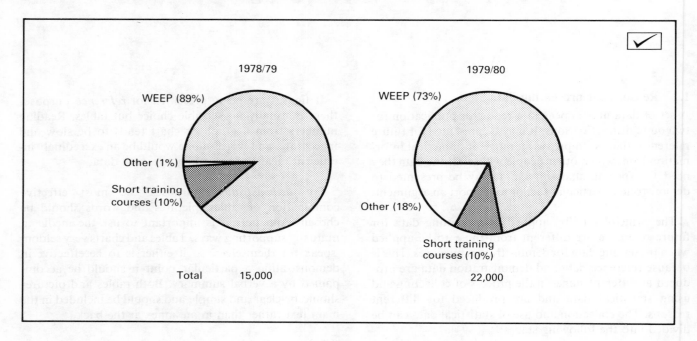

remaining 17 per cent joining a variety of different schemes.

The chart immediately highlights two qualitative relationships: the 'others' component increased markedly between 1978/79 and 1979/80, and, in both years, Work Experience on Employers' Premises accounted for most entrants to the Youth Opportunities Programme.

The significance of the change in pattern is best explained in the text which accompanied the original table.[1]

'One of the major aims in 1979/80 was to increase the range of provision available to meet the varying needs of unemployed young people. In the early days of the Programme, there was heavy reliance on the Work Experience on Employers' Premises (WEEP) element, but the table reflects the increased provision that has now been made in the other elements of YOP.'

It is usually more expensive to produce a good chart than a good table, both in terms of design time and the cost of reproduction. However, charts are more decorative and are likely to appeal more to a non-specialist audience. In choosing between tables and charts, a sensible rule is to use a table unless a particular advantage is to be gained by using a chart.

1.3   Tables and charts for demonstration. Does it work?

In considering the merits of any table or chart designed for demonstration purposes, only one question needs to

be asked, 'Does it work?' In other words, does the table or chart communicate a message to the reader clearly and simply?

Before discussing the general principles of communicating numbers effectively, it is helpful to examine examples of good and bad practice. The rest of section 1.3 is devoted to examining some examples of demonstration tables and charts. At first reading you may prefer to skip straight to section 1.4, returning to consider these examples later on.

Let us start by considering Tables 1.2 and 1.3. The first table is reproduced from *Social Trends* 12 (1982), and the accompanying commentary is given below. The second table presents the same data laid out rather differently and with a different verbal summary.

Verbal summary to Table 1.2
Trends in the destination of pupils in England and Wales in the academic year following their sixteenth birthday are shown in Table 1.2. The proportion of 16-year olds staying on at schools in the first-year sixth form rose from 26 per cent in 1973/74 (the first year after the raising of the school-leaving age) to nearly 28 per cent in 1979/80. The proportion entering full-time or sandwich non-advanced further education rose even more sharply from just under 10 per cent to over 14 per cent over the same period.

Verbal summary to Table 1.3
Over the period 1973/74 up to 1979/80 there was a fall in the percentage of 16-year olds leaving school to take up employment: in 1973/74, 61 per cent of this age group was in employment (which often included part-time day study) whereas by 1979/80 only 51 per cent of 16-year olds were in employment. Over the period there was a marked increase in the percentage proceeding to

---

[1] Manpower Services Commission Annual Report, 1979/80, (para 8.34 and Table 35)

Table 1.2 Destination of pupils attaining the statutory school-leaving age[1]

England and Wales

Destination (percentages)	1973/74	1974/75	1975/76	1976/77	1977/78	1978/79	1979/80
Staying on at school	25.9	26.1	27.5	28.3	27.6	27.4	27.8
In full-time or sandwich non-advanced further education[2]	9.7	11.5	13.6	13.6	14.1	14.3	14.1
In employment							
—with part-time day study	17.4	16.4	12.1	10.2	14.1	12.1	12.2
—with no day study[3]	44.1	41.7	38.0	37.7	34.4	38.7	38.7
Unemployed[4]	3.0	4.2	8.8	10.1	10.0	7.5	7.2
Total pupils (= 100%) (thousands)	701	723	744	746	773	801	814

[1] Destination of pupils in the academic year following attainment of statutory school-leaving age (age 16)
[2] Including a small number in advanced further education (300 in 1979/80)
[3] Produced by differencing
[4] Including unregistered estimate from the EC Labour Force Survey

Source: Department of Education and Science; Department of Employment; Welsh Office

full-time or sandwich non-advanced courses (from 10 per cent in 1973/74 to 14 per cent at the end of the period), and in the percentage unemployed (from 3 per cent to 7 per cent, although the percentage unemployed reached 10 per cent in the period 1976 to 1978). There was also a small increase in the percentage staying on at school, from 26 per cent to 28 per cent.

Table 1.3 Destination of pupils attaining the statutory school-leaving age[1]

England and Wales 1973/74 to 1979/80

	Destination (percentages)					Total pupils (100%) Thousand
	Employment		Staying on at school	In full time or sandwich non-advanced further educ[3]	Unemployed[4]	
	With no day study[2]	With part-time day study				
1973/74	44	17	26	10	3	700
1974/75	42	16	26	12	4	720
1975/76	38	12	28	14	9	740
1976/77	38	10	28	14	10	750
1977/78	34	14	28	14	10	770
1978/79	39	12	27	14	8	800
1979/80	39	12	28	14	7	810

[1] Destination of pupils in the academic year following attainment of statutory school-leaving age (age 16)
[2] Produced by differencing
[3] Including a small number in advanced further education (300 in 1979/80)
[4] Including unregistered estimate from the EC Labour Force Survey

Source: Department of Education and Science; Department of Employment; Welsh Office

Table 1.2 is attractively laid out, clearly labelled and accompanied by a verbal commentary. But most people will find the layout of Table 1.3 more helpful and its verbal summary more memorable.

There are a number of differences between the two tables, and, in each case, Table 1.2 has been altered to make it easier for the reader to assimilate the patterns in the data.

First, all entries in the second table have been heavily rounded: this allows readers to carry out mental arithmetic easily and hence to identify trends and exceptions to general trends: we can all subtract 10 from 12 and get 2, but more effort is involved in subtracting 9.7 from 11.5.

Secondly, the rows and columns have been interchanged. The reason for this is that the main patterns in the table are trends over time and it is easier to compare figures in columns than figures in rows. The decrease in the percentage who were in employment with no day study from 1973/74 to 1977/78 from 44 per cent to 34 per cent followed by a rise to 39 per cent is obvious when the figures are arranged vertically:

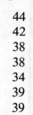

44
42
38
38
34
39
39

but less apparent when the numbers are arranged horizontally:

44 42 38 38 34 39 39

Two reasons contribute to this: numbers in columns are physically closer to each other than numbers in rows and, as in the case of drastic rounding, this layout helps the reader to carry out mental arithmetic. At school, subtraction sums were taught by putting one number on top of another and the answer underneath: few people were trained to do horizontal sums in their formative years.

A third difference is that the order of the categories has been changed so that the largest single category (those in employment) appears first, then the next largest and so on across the page. The reason for ordering columns like this is to remove arbitrary irregularity from the table. If the reader sees immediately that, in general, numbers decrease from left to right (or, in other cases, from top to bottom) any exceptions to this pattern are identified

13

quickly and are easier to remember as exceptions to a general trend. (In this table, clearly the two columns showing the percentages in employment with and without part-time day study must appear next to each other and this means that the second column is actually smaller than the third. However, the layout of the heading makes the relationship between the first two columns clear and it is comparatively easy for the reader to add together the first two entries in each row to obtain the total percentage in employment.)

The final difference between the two tables is in their accompanying commentaries. The second commentary starts by mentioning the most important trend in the table, namely the fall in the percentage in employment from 61 per cent to 51 per cent over the period covered. Since the entries in each row sum to 100 per cent, a reduction in one column must obviously be balanced by increases in one or more of the other columns, and the second part of the commentary follows logically by explaining that the 10 per cent fall in the percentage in employment was balanced by two increases of 4 per cent (in those proceeding to full-time or sandwich non-advanced further education and in those unemployed) and an increase of 2 per cent in the numbers staying on at school. Thus this commentary starts by highlighting the main pattern in the data and proceeds naturally to

mention related but less dramatic trends. By contrast, the first commentary mentions only two minor trends, the change in the proportion staying on at school and the change in the proportion entering full-time or sandwich non-advanced further education. In order to assimilate the major patterns in this table, the reader has to do further analysis for himself.

Just as a good table makes the patterns and exceptions obvious at a glance, so a good chart instantly communicates a clear message to the reader. By contrast a poor chart leaves the reader puzzling over its interpretation.

Figures 1.2 and 1.3 with accompanying commentaries are both taken from *Social Trends* 12 (1982).

Verbal summary to Figure 1.2
Figure 1.2 shows the number of days of certified incapacity for sickness and invalidity benefit. In 1979–80, 83 million such days were lost by females and 276 million days by males.

Verbal summary to Figure 1.3
The number of industrial stoppages and working days lost each year is shown in Figure 1.3. Wage disputes were the main reason for the increase in the number of working days lost in the 1970s. The residual category

Figure 1.2   Days of certified incapacity for sickness and invalidity benefits[1]: by age and sex

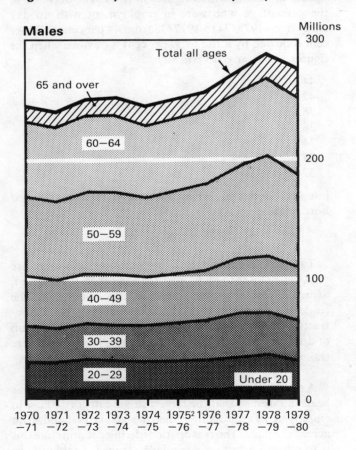

¹ See Appendix, Part 4: Certified incapacity for sickness and invalidity benefits.
  Data are shown for the 12 months starting on the first Monday in June.
² Data for 1975-76 are not available.

*Source: Department of
Health and Social Security*

Figure 1.3 Industrial disputes—working days lost: by cause; and number of stoppages

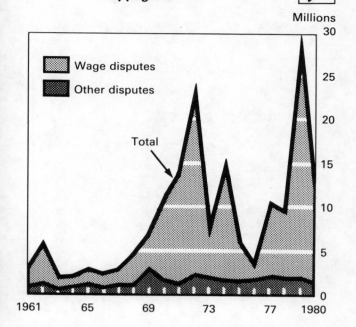

Millions

- Wage disputes
- Other disputes

Total

1961 65 69 73 77 1980

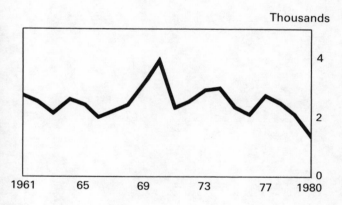

Thousands

1961 65 69 73 77 1980

'other disputes' covers disputes on hours of work, redundancy, trade union matters, work allocation, or disciplinary matters.

Stoppages in 1980 (1,330) were the lowest for nearly 40 years. There has, however, been an increase in the average number of working days lost per stoppage, reflecting the increase in the number of large scale disputes.

The number of working days lost through industrial stoppages in 1980, at 12 million, was just below the annual average over the previous decade. The national steel strike in the first three months of 1980 accounted for nearly 9 million working days lost—three-quarters of the total for the year.

The essential difference between Figures 1.2 and 1.3 is that one has a clear story to tell and the other has not. In Figure 1.2 a great deal of data has been carefully converted to graphical form, but what pattern emerges? Because the numbers for each age group are shown

cumulatively it is very difficult to identify changes over time in any single group (other than the under-20 category which is plotted from a horizontal base line), and because different scales have been used for males and females it is virtually impossible to make comparisons between the sexes. Instead of helpfully directing the reader's attention to any memorable patterns in the data, the commentary states baldly:

'Figure 1.2 shows the number of days certified incapacity for sickness and invalidity benefit. In 1979–80, 83 million such days were lost by females and 276 million days by males.'

By contrast Figure 1.3 is much less ambitious yet much more effective. Comparatively little data is shown in the two component charts, but clear patterns are immediately obvious and these are highlighted in the accompanying text. The charts have a number of messages to communicate:

—Wage disputes were the main reason for the increase in the number of working days lost;

—Stoppages in 1980 were the lowest for a long time;

—The number of working days lost through industrial stoppages in 1980 was 12 million and this was just below the annual average over the previous decade.

In Figure 1.3, the graphs and the commentary work together to communicate information to the reader clearly and memorably.

1.4 General principles for demonstration tables and charts

The design of demonstration tables is discussed in detail in Chapter 4, and Chapters 6 and 7 are devoted to graphical presentation. In both cases, the prime objective is to devise a method of communicating patterns in the original data as effectively as possible.

In the case of tables this involves displaying numbers in a way which allows readers to compare numbers with each other as easily as possible (using mental arithmetic if necessary) and in a way which emphasises any general patterns in the data (such as a decrease in magnitude from left to right across the table or from the top to the bottom of the table): exceptions to a general pattern are then readily identifiable.

In the case of charts the same general principle leads to the use of simple, familiar forms of chart, like pie charts, bar charts of various sorts, and line graphs showing how one or more measurements varied over time. It also limits the amount of information that can be encapsulated effectively in a single chart to two or three clear patterns.

Both tables and charts should always be accompanied by a verbal summary which repeats, and therefore reinforces, the message illustrated. This summary should be close enough to the table or chart for readers to glance back from the summary to the table or chart to check their understanding of each point mentioned: the 'Ah yes, I see' reaction should be encouraged.

1.5 Reference tables
Are they easy to use?

The design of reference tables is discussed in detail in Chapter 5. The objective is simply to provide the data required in a form that the reader will find easy to use. This involves giving as much detail as can practically be provided and including helpful totals and subtotals. It will also frequently involve choosing categories and an order of tabulation which are consistent with related tables. It is important to include complete definitions of the terms used in the table and to note any changes in defnition of terms or in methods of data collection which may affect the validity of comparisons between different entries in the tables.

1.6 Accuracy

In reference tables it is important to record the intrinsic accuracy of the data. This can be done explicitly by giving ranges or standard errors (if appropriate to the expected user) or implicitly by rounding all figures appropriately.

In demonstration tables, most figures should be presented in rounded form. Inevitably this involves some loss of accuracy, but it is important to realise, in this context, that if an apparent pattern in the data vanishes when the figures are rounded appropriately, it is doubtful if the pattern is robust enough to form a basis for decision-making. This point is discussed in more detail in Chapter 4.

1.7 Summary of general principles

Reference tables should be easy to use speedily and accurately. Demonstration tables and charts should communicate a small number of messages clearly and simply; they always need an accompanying verbal summary.

Chapter 2: Summary of recommendations

2.1 Overview

This chapter presents, in condensed form, the guidelines which are discussed and illustrated in the rest of the book. It may be read alone but, to understand the reasons behind each recommendation and to see the effect of following specific guidelines, the appropriate chapter should be read in full.

2.2 Choice of medium for presentation

The first question to be resolved in presenting numbers is whether to use words, charts or tables. Few people would argue in favour of using words alone to present more than two or three isolated numbers, and so the usual choice is between tables and charts.

Most books and articles on the subject of presenting data fall into one of two categories: they either assume a commitment to graphical presentation and contain details of many ingenious graphical devices, or else they adopt a non-committed standpoint on whether tables or charts should be used to present statistical data.

The introductory chapter promised some positive guidance and this chapter summarises the recommendations discussed in the rest of the book. If these guidelines are followed carefully, even the most inexperienced presenter should produce effective tables and charts.

2.3 Demonstration or reference?

When planning to display any set of data, the first question to ask is

'Are these data being presented to demonstrate a particular point or are they for future reference?'

If the answer is 'for future reference', a table should be used, and a summary of the guidelines for constructing reference tables will be found in section 2.9. Otherwise, read on. (Please try to avoid saying 'Both': no chart or table can serve both purposes well. This point is elaborated in Chapter 3.)

2.4 For demonstration, charts or tables? Write the verbal summary first

Charts tend to be time-consuming and expensive to produce, so it is important to know when a chart is likely to illustrate the patterns in a set of data more effectively than a table. Table 2.1 on page 19 lists the main advantages and disadvantages of charts, with the suggestion that a chart should be used only when the advantages outweigh the disadvantages. However, an initial assessment of whether or not a chart is likely to be more effective than a table can be made by considering what sort of patterns the data exhibit.

Since neither charts nor tables ever 'speak for themselves', in order to communicate a message effectively either must be accompanied by a verbal summary. And the final choice of whether to use a chart or a table is best made *after* writing the verbal summary.

The summary should be a clear, concise statement of how the data contribute to the subject of the report, consisting of no more than three or four separate points each clearly supported by the accompanying table or chart. Chapter 8 offers guidance on how to write a verbal summary of a table or chart. Before deciding on the verbal summary, a number of tables and rough graphs will probably have been constructed and the words finally chosen will often suggest the appropriate form of data display. On occasions both a table and a graph may be needed.

2.5 Tables for specific amounts

If the verbal summary emphasises *specific amounts*, then a table should be used. For example, the following is an extract from a verbal summary:

'The increase in the number of Soviet military advisers in Third World countries from 3,700 in 1965 (in 14 such countries) to about 15,000 in 1982 (in 31 such countries) is less well-known but . . .'

This clause stresses both the number of Soviet military advisers and the number of Third World countries in which they are based. Here the summary calls for an accompanying table which repeats the numbers quoted in the text and shows corresponding numbers for some intermediate years.

2.6 Charts for comparisons

Alternatively, if the verbal summary emphasises broad *comparisons*, such as:

'In social classes C1, C2, D and E, the *Sun* is the most popular daily newspaper; in social classes A and B the *Daily Telegraph* is most popular . . .'

or if it includes a number of *non-specific* quantitative statements, such as 'the trend levelled off'; 'the percentage of homes owning a colour television set increased dramatically', then a chart may be most effective. Charts do, however, require more space and they can be more difficult to prepare. Section 2.11 contains a more detailed analysis of what sort of chart is likely to be most appropriate in specific circumstances.

2.7 Which numbers to present

Demonstration tables may contain either the original data, often heavily rounded for ease of comprehension, or derived statistics such as percentages, indices or ratios. The choice of which numbers to present in the final table is best made after analysing the data in a number of different ways and establishing the main patterns and exceptions. The numbers which best illustrate these patterns and exceptions are the ones to present. For example, if changes in the composition of a total are emphasised, *percentages* should usually appear in the final table; if growth patterns in several different measurements are to be compared over a number of years, comparisons may be easier to make if all the measurements are expressed as *indices*, based on a common year whose value is taken as 100 in each of the series being examined; if two related measurements (such as total population and national spending on health) have grown at different rates, it may be informative to tabulate their *ratio* (£s per head); if an annual total results from an increase (such as births) and a decrease (deaths), it may be appropriate to tabulate the *differences* between these measurements as a net annual gain or loss; in some cases, valid comparisons can only be made if original numbers of observations are re-expressed as *rates per number at risk* (for example, births in different areas would be related to the number of women of child-bearing age; accidents in different sports might be related to the number of 'player hours' or 'player sessions'); and growth in any financial measurement can usually be assessed correctly only after it has been expressed *at constant prices*.

2.8 Guidelines for demonstration tables

The criterion for a good demonstration table is that 'the patterns and exceptions should be obvious at a glance, at least once one knows what they are'. To this aim the following basic guidelines are proposed:

a. Round numbers to two effective digits: this facilitates mental arithmetic and makes the numbers easier to remember.

b. Put the numbers most likely to be compared with each other in columns rather than rows: like guideline a., this makes quick comparisons easier since the numbers are physically closer to each other in columns than in rows.

c. Arrange columns and rows, where possible, in some natural order of size; if there is no obvious natural order, arrange columns and/or rows in decreasing order of their averages. This helps to impose as much visible structure as possible on the table, and exceptions to a general decreasing pattern are imediately highlighted.

A corollary to guideline c. is that wherever possible large numbers should appear at the top of columns, again to aid mental arithmetic (we subtract by putting the larger number on top), and to highlight the average patterns in the data. But an exception to this guideline must be made in government statistical publications where *time* trends are displayed in columns: in virtually all official statistical tables the most recent time period is shown at the *bottom* of columns, and this practice should be followed for readers accustomed to this layout.

d. Give row and/or column averages or totals as a focus: they provide a structure which helps the reader to see the 'average' relationships between rows and/or columns, and this in turn makes it easier to note and remember deviations from the average pattern.

e. Use layout to guide the eye: columns and rows in the body of the table should be regularly spaced and close together; additional blank spaces may be used to separate rows or columns showing averages or totals from the rest of the table; if the table contains more than six rows, an artificial break should be used to divide columns into blocks of four or five items for ease of reading (but demonstration tables should seldom be big enough to require such an artificial break).

f. Accompany each table with a clear verbal summary, positioned close to the table and referring directly to numbers which appear in the table: this allows the reader to check his understanding of how the table supports the text, thus reinforcing the message and making it more memorable. No more than three or four distinct points should be made and these should summarise the *main patterns* in the data. Any necessary explanations of how the data were collected or collated should be shown separately.

A final guideline in constructing demonstration tables is that they should be small. It is better to include three or four compact tables, each illustrating one or two points succinctly, rather than to construct a single large table which is then referred to in text covering a number of paragraphs or pages. Small tables are easier to position close to their verbal summary, and are easier to include in the main report. They are also more digestible than large tables.

The standard layout, described in Chapter 5 and summarised in points a. to d. below, should be used for the edgings of the table.

2.9 Guidelines for reference tables

The criterion for a good reference table is that it should be easy to use: it should be clearly and legibly laid out and the data should be precisely defined. The first four guidelines, on layout, apply to demonstration tables as well as reference tables.

a. All the information necessary to interpret the table should appear in the table: this can be achieved by using the standard layout (described in Chapter 5) to provide clear information on:

—the kind of objects enumerated

—the factors by which they are tabulated

—the units of measurement

—the geographical coverage

—the time period covered

—the source of the data.

b. Use horizontal lines and blank space to guide the eye. Columns and rows should be regularly spaced and as close together as is consistent with clarity. Horizontal lines may be used to separate column headings and a total row from the body of the table. Average rows and/or columns and total rows and columns should be separated from the rest of the table by extra blank space.

Long columns should be divided into blocks of about five entries by an artificial break (a blank row).

Columns in the body of a table should not be separated by vertical lines.

c. Tables should be printed upright on the page: if a large reference table extends over two pages, consider repeating the row headings at the right hand side.

d. Detailed guidelines on:

—choice of typeface and use of capital letters

—table titles and row and column headings

—numbering of tables

—punctuation marks for large numbers

—layout and use of footnotes

are given on pages 28 to 31 of Chapter 3. The guidelies proposed there are detailed and concise.

e. Choice of categories to show separately: where possible use a standard set (for example, the Standard Industrial Classification): if in doubt consult the users of the tables. Be consistent from one year to the next and throughout sets of tables on related topics.

f. Put in columns the items which will be compared with each other most frequently: it is easier to scan up and down a column than along a row.

g. Indication of probable size of error: consider the needs of the likely users of the tables. For general use the figures may be rounded so that they are accurate to the last non-zero digit shown. For more specialised use it is better to give the probable size of errors explicitly either by:

—attaching quality labels to each row or column (for example, label A might indicate ranges of error of less than 1 per cent, label B might indicate possible errors of between 1 per cent and 10 per cent, and so on);

—or giving the standard error associated with each entry separately (though this can lead to a congested table).

h. Include full and clear footnotes, using covering notes conservatively (they may not be read: footnotes are more obvious).

i. Include a complete set of definitions of the terms used in the table and also a key to abbreviations and special symbols.

j. Make sure that a clear account of the methods used to collect the data and compile the table is readily available for technical readers.

k. Commentary on reference tables is not normally necessary. If circumstances call for such a commentary it is important to ensure that it is impartial and that the reader will have no difficulty in linking all points made in the commentary with the appropriate source table.

2.10 When to use charts.
Demonstration only, never for reference

Charts, graphs and diagrams seldom have a role to play in presenting data for *reference* purposes. Reading numbers from a chart tends to be slow and imprecise. However, a carefully designed chart may well be more effective than a table when used for *demonstration*.

Charts have a number of advantages and disadvantages; you should consider using a chart rather than a table when the pluses outweigh the minuses in the following table.

Table 2.1 Charts: plus points and minus points

+	−
Good for communicating non-specific quantitative comparisons	Poor for communicating specific amounts
Attractive to look at: make a report look more interesting	Time-consuming and expensive to design and execute well
Likely to appeal more to a general audience than a table of figures	May not be properly assimilated: people need to be trained to interpret any but the simplest sort of graph
General time trends can be shown and compared more effectively using line graphs than using tables	May not be possible to show related trends on the same graph if the orders of magnitude are different

2.11 What sort of chart?

A bewildering variety of statistical charts and diagrams have been devised, some requiring considerable skill to interpret correctly. Since it is unrealistic to expect a general reader to pore over an unfamiliar chart in order to decode its conventions, only simple, straightforward charts should be used when writing for non-specialists.

The main strength of charts is their ability to depict

such comparisons as changes in the composition of a total, changes over time in one or more measurements and changes in the ranked order of a number of measurements. The following table lists the most frequently encountered comparisons and the charts which may be used to illustrate each.

Table 2.2 Which chart for which comparison?

Type of comparison	Possible charts
1. Components as parts of a whole	Pie chart. Component bar chart.
2. Change in composition of a total	Pie charts. Component bar charts (vertical or horizontal bars).
3. Sizes of related measurements	Grouped bar charts (vertical or horizontal bars). Possibly isotypes.
3a. Frequency with which different measurements occurred	Bar charts. Histograms.
4. Change over time in one or more related measurements	Line graph. Column bar charts. Possibly isotypes.
5. Change in ranked order of a set of measurements	Paired or grouped bar charts.
6. Correlation between two sets of measurements	Back-to-back bar charts. Scatter diagrams.

General guidelines which apply to all charts are:

a. Each chart should include a clear definition of

—the kind of objects represented

—the units and scale of measurement used

—the geographical coverage

—the time period covered

—the source of data.

b. The chart should be easy to read: it should be upright on the page and no bigger than necessary for clarity; lines and sections of the chart should be labelled directly rather than via a separate key and, if shadings are used, they should be easy to distinguish from each other.

c. The chart should be easy to interpret: this means that the conventions used in constructing the chart should be self-explanatory and not distracting (for example, bars of different widths *and* different heights should not be used): it also means that each chart should be accompanied by a clear verbal summary of its main points.

d. Prefer to use two or three small, simple graphs to build up a story rather than put too much on a single chart.

e. If the use of a specialist graph seems to be justified because of the clarity with which it alone can illustrate a particular point, take time and care to train the reader in how to interpret its conventions: give a simple example of how a familiar measurement would appear on it (for example, regular compound interest on a logarithmic scale).

2.12 The role of words in communicating numbers

Words alone should *never* be used to communicate more than two or three simple numbers. But a verbal summary should accompany *every* table and chart used for demonstration purposes. Since most people find numbers alienating and uninteresting, the prose used in summarising quantitative information should be as lucid and readable as possible.

In writing a verbal summary of a table or chart, include only three or four major patterns displayed by the data; do not include a blow-by-blow account of every row or column and avoid explanations about how the data are defined or tabulated unless it is essential to point out that an apparently dramatic pattern is attributable to a break in the series. (Such information should, of course, appear clearly as a footnote to the table or chart itself.)

In writing a longer report about a statistical investigation or in producing a summarising commentary on a set of reference tables, aim to make the report as readable as possible. Measures which can help to achieve this are:

—putting the conclusions first

—using short sentences with as few qualifying clauses as possible

—avoiding technical terms as far as possible

—using sub-headings freely

—relating the report to familiar information (by using analogies, or by relating the data to items of general knowledge, for example, 'the year of the miners' strike').

Finally, the most important rule in communicating quantitative information is to THINK CLEARLY. If you know exactly what your data say you will have little difficulty in communicating the message effectively.

Chapter 3: Structure and style of tables

3.1 Introduction

Once you have decided to present a set of data in a table a number of details must be considered. For example, you must decide what title to give the table, where to show the units of measurement, how much space to leave between rows and columns, what footnotes to include and what typeface to use. These are decisions on how the table is to be *structured* and on its *style* of presentation.

This chapter discusses the structure and style of tables. The guidelines offered apply to both reference and demonstration tables.

3.2 Structure

Just as there are rules to be obeyed if a passage of prose is to communicate a clear and unambiguous message, so there are rules associated with constructing a table of numbers. In the first case, the prose must be grammatically correct; in the second case, the table must be correctly structured.

Fortunately the rules associated with table structure are fewer and simpler than the rules of grammar. They concern the *definition* and *layout* of tables and they can be set out in a few pages.

Most experienced presenters of data acquire some basic knowledge of table structure by following good examples or by trial and error. Although this chapter may be of interest to such presenters, it is really designed to help inexperienced presenters produce clear professional tables *every time*.

Table definition

Every table should contain a clear and complete explanation of the figures presented. The reader should be left in no doubt about:

1. the kind of events, objects, people etc which are enumerated

2. the factors by which the data are categorised

3. the units used, eg absolute numbers, thousands, rates, percentages

4. the geographical coverage

5. the time period

6. the source of the data (if the document in which the table appears is not itself a primary source).

A common mistake is to rely on separate text to convey some of this information. This can lead to figures in a table being taken out of context and misinterpreted. The standard arrangement, used in most Government Statistical Service publications and papers, is to include 1. and 2. in the table title, to state 3., 4. and 5. immediately above the main body of the table and 6. underneath, like this:

Table 3.1A   [Kind of objects, etc]: by [classifying factor or factors]

Geographical and time coverage   *Units*

[Data, in rows and columns]

Source

For example:

Table 3.1B   Deaths from fires in dwellings: by source of ignition

United Kingdom, 1970 to 1976   *Number of deaths*

(Data in rows and columns)

Source: Home Office

Footnotes should be used to record any changes in definition or changes in the method of data collection which may invalidate comparisons between individual rows or columns within the table.

This arrangement works well in practice and is strongly recommended. It provides a useful discipline and gives the top of the table a neat, professional look. The standard layout need not be used always, but presenters should only depart from it with good reason.

Where a table consists of columns measured in different units, each column must be headed with the appropriate units and the 'units' position at the top right-hand corner left blank. See for example Table 3.2.

Standard layout for two-way tables

Many tables consist of data tabulated according to two (or more) factors: for example, the number of fires in dwellings can be grouped into distinct categories according to the source of ignition and then further sub-divided to show the material first ignited within each source of ignition. Data displayed like this are said to be *cross-tabulated*: this means that one margin of the table (say, the row headings) records the source of ignition, and the other margin (the column headings) records the material first ignited. Alternatively, the table may be

Table 3.2: Trends in employment and output of the *Daily Telegraph*

1972 to 1976

Year	Number of employees	Pages per day (average number)	Circulation (millions)	Pages produced per day (millions)	Pages per employee on payroll per day
1972	2,728	31.5	1.43	45.0	16,512
1973	2,704	32.9	1.42	46.7	17,277
1974	2,692	30.9	1.40	43.3	16,070
1975	2,724	27.9	1.33	37.1	13,622
1976	2,677	28.1	1.30	36.5	13,646

Sources: Daily Telegraph, Royal Commission on the Press (Final Report 1977, Cmnd. 6810) and Audit Bureau of Circulation

referred to as a *two-way table* to indicate that it displays data analysed by two separate factors.

The recommended layout for a two-way table is given at Table 3.3 below.

Table 3.3: [Kind of objects, etc]: by [factor A and factor B]

Geographical and time coverage   *Units*

Factor A	Factor B			Total
	Category B1	Category B2	etc	
Category A1				
Category A2				
etc				
Total				

Source

The complete table on fire deaths (Table 3.4) will serve as an example.

Table 3.4 Deaths from fires in dwellings: by source of ignition and material first ignited

United Kingdom, 1976   *Number of deaths*

Source of ignition	Material first ignited			Total
	Upholstery or other textiles	Other	Unknown	
Smoking	153	6	16	175
Space heating	120	30	14	164
Other	101	51	35	187
Unknown	12	8	144	164
Total	386	95	209	690

Source: Home Office

As before, there is no need to follow the standard layout slavishly where there is a well-considered reason for using an alternative structure.

Development of the standard layout for use in more complex tables

The standard layout can be extended to display data analysed according to three or even more factors, but this is seldom advisable. Few people will be able to see the overall pattern in a demonstration table if the data are analysed according to three different factors, and mistakes are likely to occur in using such reference tables.

However, where there are special reasons to justify a three-way table, the standard layout can be adapted in a logical fashion. For example, the table of fire deaths could be expanded to include cause of death as a third factor (see Table 3.5).

When some figures in a table represent one kind of entity and others represent different kinds, it is good practice to divide the table into separate, clearly labelled parts, each part containing only one kind of object. This can be done using the same type of structure as in Table 3.5. For example, Table 3.6 shows the number of divorcing couples in the first section of the table, the number of children of divorcing couples in the second section, and has a final one-line section showing the average number of children per divorcing couple.

The three sections are separated from each other by blank spaces and each section is clearly labelled to identify the entity being tabulated.

Table 3.5 Deaths from fires in dwellings: by cause of death, source of ignition and material first ignited ✓

United Kingdom, 1976

Number of deaths

Cause of death	Source of ignition	Material first ignited			Total
		Upholstery or other textiles	Other	Unknown	
Overcome by gas or smoke	Smoking	110	1	13	124
	Space heating	58	18	10	86
	Other	53	29	19	101
	Unknown	8	2	79	89
	Total	229	50	121	400
Burns	Smoking	36	5	1	42
	Space heating	56	8	2	66
	Other	43	14	10	67
	Unknown	4	5	46	55
	Total	139	32	59	230
Other and unknown	Smoking	7	–	2	9
	Space heating	6	4	2	12
	Other	5	8	6	19
	Unknown	–	1	19	20
	Total	18	13	29	60
Total	Smoking	153	6	16	175
	Space heating	120	30	14	164
	Other	101	51	35	187
	Unknown	12	8	144	164
	Total	386	95	209	690

Source: Home Office

The rule for dividing the table into parts applies just as strongly when the two entities concerned are numbers and percentages or numbers and rates. For example, Table 3.7 gives details of marriages in selected years from 1966 to 1978. The top part of the table records the *number* of marriages solemnised in different manners and the lower part the *percentage* of marriages solemnised in Register Offices in different geographical areas.

Demonstration tables should seldom contain different entities. Where it is desirable to tabulate two sets of related data, or to show both percentages and original numbers, this will often be done more effectively by using separate tables than by expanding a single table.

Table 3.6 Divorcing couples and children of divorcing couples: by age of children ✓

England and Wales, 1971 to 1976

	1971	1972	1973	1974	1975	1976
Divorcing couples			Thousands of couples			
With						
—No children under 16	32	52	42	45	47	49
—1 child under 16	17	27	24	26	28	30
—2 children under 16	15	24	24	26	28	31
—3 or more children under 16	10	16	16	16	17	18
Total divorcing couples	74	119	106	114	121	127
Children under 16 of divorcing couples			Thousands of children			
Ages of children						
—under 5	21	30	30	32	33	34
—5 but under 11	41	64	62	65	69	71
—11 but under 16	21	36	36	39	43	47
Total children under 16 of divorcing couples	82	131	127	135	145	152
Average number of children under 16 per divorcing couple			Number of children			
	1.11	1.10	1.20	1.19	1.20	1.20

Source: Office of Population Censuses and Surveys

Table 3.7 Marriages: by manner of solemnisation

Great Britain ☑

Manner of solemnisation	1966	1971	1976	1977	1978
					Thousands
Church of England/Church in Wales	175	160	120	117	120
Church of Scotland	22	20	16	15	15
Roman Catholic Church	51	48	34	34	
Other Christian	38	37	31	32	70
Jews, other non-Christian	2	2	1	1	
Register Office	138	180	193	195	200
Total Marriages	426	447	396	394	405
Marriages solemnised in Register Offices as percentages of all marriages					Percentages
England and Wales	33	41	50	51	51
Scotland	25	31	38	38	40
Great Britain	32	40	49	49	49

Source: OPCS; Register Office for Scotland

Sub-totals

Sub-totals pose an awkward presentational problem. Take, as a simple example, a table showing number of journeys to work, analysed by method of transport and length of journey. There are four main categories of method of transport—'private vehicle', 'public transport', 'walking' and 'other'—but 'private vehicle' contains the sub-categories 'car', 'motor cycle' and 'bicycle', and 'public transport' contains the sub-categories 'bus' and 'train'. How do you construct an easy to use table showing all the figures for categories and sub-categories? Clearly it would not be satisfactory to simply list all the figures like

Method of transport	Length of journey	
	Under 3 miles	3 to 6 miles
Private vehicle	146	etc
Car	83	
Motor cycle	37	
Bicycle	26	
Public transport	122	
Bus	94	
Train	28	
Walking	66	
Other	11	
Total	345	

A presentational device is needed to show that the figures 146, 122, 66, 11 and 345 have a different status from the others. If the table is being printed, the lines containing these figures can be entered in bold print. This method, especially when used together with appropriate indentation and row spacing, can be highly effective. But when only simple typescript is available the problem is more difficult and we have not found any completely

satisfactory method. Extra ruled lines can sometimes be used to emphasise the status of sub-totals (see, for example, Table 3.16) but this only works if all the main categories have sub-categories.

In the example here the 'walking' and 'other' categories are not sub-divided, and it is better to employ just row spacing and indentation:

Method of transport	Length of journey	
	Under 3 miles	3 to 6 miles
Private vehicle	146	etc
Car	83	
Motor cycle	37	
Bicycle	26	
Public transport	122	
Bus	94	
Train	28	
Walking	66	
Other	11	
Total	345	

This is not entirely satisfactory, but it is probably the best that can be done when only one typeface is available. In a printed table, the five main rows could be entered in bold print and there would be no need to inset the sub-category figures; it would however still be helpful to inset the row captions and to take care with row spacing.

Summary of rules

1. State the entities tabulated and the factors by which they have been grouped in the table title.

2. Use the edges of the table to include clear information on:

   —units of measurement

   —geographical and time coverage

   —source of data.

3. If different entities are included in the same table, separate the table into two or more parts, labelling each part clearly to show what entities are tabulated in it and what are the units of measurement. Consider dividing tables which contain two or more entities into separate tables.

4. Indent row headings to indicate entries which are components of totals and sub-totals.

3.3 Applying the rules on structure

Simple as they are, the rules suggested for the construction of statistical tables represent the essence of good professional practice. The reason for most people producing

Table 3.8 Further Education under 18 Day Release Scheme, years ending 31 July 1977-1980

	1979/80	1978/79	1977/78
— Total number of staff under 18 recruited between 1 August 1979 and 31 July 1980 eligible for further education day release.	9,058	13,038	13,131
— Total number of eligible staff who were enrolled on courses commencing between 1 August 1979 and 31 July 1980 (including those on block release or engaged on correspondence courses in lieu).	2,282 (25.2%)	4,225 (32.4%)	5,412 (41.2%)
— Total number of eligible staff (ie staff on vocational training) who were deferred to begin day release courses commencing after 31 July 1980	4,219 (46.6%)	5,509 (42.3%)	4,290 (32.7%)
— Total number of eligible staff who:			
A. Remained in the Service and abstained from enrolment	2,211	2,689	3,429
B. Left the Service before enrolment	346 (28.2%)	615 (25.3%)	(26.1%)
— Breakdown of type of course on which eligible staff were enrolled:			
A. Civil Service Limited Competition and general clerical development courses	69 (3.02%)	244 (5.8%)	245 (4.5%)
B. General Certificate of Education	1,282 (56.1%)	2,188 (51.7%)	2,917 (53.9%)
C. COS/ONC/HNC or Business Education Council Successors	598 (26.2%)	1,299 (30.7%)	1,923 (35.5%)
D. Others	333 (14.6%)	494 (11.7%)	327 (6.0%)

Explanatory Note
Sections B, C and D are a breakdown of the total shown in Section A and therefore the numbers provided in each of these sections must add up to the number provided in Section A. Sections A to D and Section E which is a breakdown of Section B, refer to those staff eligible for day release who were recruited between 1 August 1979 and 31 July 1980.

tables of a standard far below that expected of a good professional statistician is that they do not know these simple rules. Consider, for example, Table 3.8, which is taken from an official report and compiled by someone who clearly knew his material well, but failed to present it effectively.

Unless you were already familiar with the detailed workings of the Under 18 Day Release Scheme, spotted immediately the vital but rather obscure explanatory note, and managed to identify at once the unlabelled sections A, B, C, D and E (which are different from those actually given such labels), the chances are you would spend many minutes trying to get a proper grasp of what these figures represent. The compiler has not even made it clear that the figures in each column represent people recruited between 1 August and 31 July in the year shown. But suppose he had applied the simple rules summarised in section 3.3. He would have followed a chain of thought similar to this:

Q1. What am I enumerating?

A1. Staff eligible and applying for the Under 18 Day Release Scheme.

Q2. By what factors are the data analysed?

A2. This needs some thought: in fact there are several. One factor is the 'year' of recruitment, with the 'year' being measured from August to July rather than January to December. The other main factor is what may be termed 'enrolment activity': whether staff applied for enrolment or not and, if they applied, whether they were (1) enrolled during the 'year' they were recruited or (2) deferred or left before taking up a place. A further factor is the type of course, but this only applies to those enrolled in the year of recruitment.

Q3. Following on from A2, would it not be sensible to put the analysis of numbers of people actually enrolled, by type of course and 'year' of recruitment, in a separate table?

A3. Yes it would; each of the two resulting tables would then provide the same degree of analysis for all its data.

Q4. What is the purpose of putting percentages in the table(s)?

A4. To enable the reader to compare proportions, year by year. This purpose would be better served if the percentages were tabulated separately from the numbers.

Had he taken this systematic approach, and known the standard layout, the person who compiled Table 3.8 would probably have produced instead something resembling the clearer, more professional Tables 3.9 and 3.10. The important point demonstrated by this example is that thoroughly bad tables should never be produced. To present the same data in a form which shows proper respect and consideration for the reader only requires thought and the application of a few simple rules.

There is, of course, no one 'right table' for any given set of data. There may be several different structures that will present the data clearly and accurately, but using the rules discussed in this chapter will ensure that one of these tables is produced. However, just as it is worth drafting and revising a passage of writing, it is worth trying several different designs of table in order to decide which one displays the data most effectively.

3.4 Style

Style plays the same role in statistical tables as in language. The first requirement for a table is that its basic structure (grammar) should be correct; the second is that its style should be effective for the task in hand. A particular type of style may be good in one application but bad in another, depending on information in the table and customers who are to use it. The test of good style is its success in getting information across correctly,

Table 3.9 Staff eligible for Under 18 Day Release Scheme: by enrolment activity and year of recruitment

Civil Service in Great Britain   *Numbers and percentages*

| | Year in which recruited (1 August to 31 July) | Applied for enrolment | | | Abstained from enrolment | Total |
		Enrolled during period[1]	Deferred to later period	Left before enrolment		
Number of Staff	1977/78	5,412	4,290	....3,429....		13,131
	1978/79	4,225	5,509	615	2,689	13,038
	1979/80	2,282	4,219	346	2,211	9,058
Percentage	1978/78	41	33	....26....		100
	1978/79	32	42	5	21	100
	1979/80	25	47	4	24	100

[1] Includes enrolments to correspondence courses or on block release.

Table 3.10 Staff enrolled on courses during period in which recruited[1]: by type of course and date of recruitment

Civil Service in Great Britain   *Numbers and percentages*

| | Period in which recruited (1 August to 31 July) | Type of course | | | | Total |
		GCE	BEC[2]	Civil Service internal[3]	Other	
Number of Staff	1977/78	2,917	1,923	245	327	5,412
	1978/79	2,188	1,299	244	494	4,225
	1980/81	1,288	598	69	333	2,282
Percentage[4]	1977/78	54	36	5	6	100
	1978/79	52	31	6	12	100
	1980/81	56	26	3	15	100

[1] Includes enrolments to correspondence courses or on block release.
[2] Includes COS, ONC and HNC courses.
[3] Limited competition and general clerical development courses.
[4] Percentages are rounded independently and do not necessarily sum to 100.

economically and, where necessary, powerfully. If a table works well, its style is, by definition, good. Nevertheless rules of style do exist. They do not have the self-evident correctness of the rules of structure and the reasons for proposing them vary: some have been checked under controlled experimental conditions while others owe their importance to the fact that they have been found, in practice, to work well most of the time. They are best thought of as guidelines to follow unless there is a sound reason for not doing so.

Before discussing these rules, let us clear the ground by considering in more detail the two distinct roles that tables can play in statistical presentation.

Reference tables and demonstration tables
In Chapter 1 we explained that statistical tables can serve two quite different purposes:

1. provision of data for reference (essentially a storage function)

2. demonstration of particular facts or situations (essentially a means of influencing decisions).

The main aim in 1. is to provide figures which the user will find easy to look up and understand; in 2. it is to get particular messages across to the reader.

The style of a table will depend strongly on whether it is for reference or for demonstration purposes. For example, if a table giving population of counties in England were intended for reference, the counties would be best listed in alphabetical order whereas, for demonstration purposes they would probably be better shown in some other order, such as population size.

Wherever possible, tables should be dedicated wholly to purpose 1. or wholly to purpose 2. and the style of presentation should be chosen accordingly. Any attempt to serve both purposes in the same table should be resisted: it is most unlikely that the same table will serve both purposes well and the danger of ending up with a table which serves neither properly should be borne in mind. If some such attempt *must* be made, decide which is the *primary* role of the table and design the table to carry out that role effectively. Then consider how the table would have to be changed in order to fulfil the secondary role effectively, for example, by rounding all the numbers in a reference table for demonstration purposes. At that point a compromise may have to be reached.

Factors affecting 'style' are:

—layout: spacing of rows and columns and use of ruled lines

—choice of typeface and use of capital letters

—numbering of tables

—punctuation marks for large numbers

—footnotes

—rounding.

Layout
A very common mistake is to allow the spacing of rows and columns to be determined chiefly by the length of row captions and column headings. This usually results in a table that is untidy and difficult to read. It is nearly always possible to shorten or rearrange the captions and headings so that they do not interfere with spacing.

Another common mistake is to have columns too widely separated. This usually arises because typists and typesetters, unless otherwise instructed, will space the columns out to cover the full width of the page. Irrespective of the number of columns, spaces between them generally need only be just big enough to prevent confusion of figures in one column with those in another and to allow the eye and mind to identify figures in individual columns as distinct groups. Somewhere between five and eight character units of blank space between figures in neighbouring columns is generally about right. Wider spacing should only be used to separate distinct sets of columns, for example, to separate the total column from its component columns.

Much of what has been said for columns also applies to rows. In general, rows of figures should be spaced no further apart than normal rows of text. Wider spacing should only be used to separate distinct sets of rows or to create artificial 'breaks' in large blocks of consecutive rows. These breaks should be inserted after every fourth or fifth row. This simple device enables the eye to identify a row much more quickly and reliably than it could if all the rows were equally spaced.

A typical spacing pattern for the numbers in a medium-size two-way table would therefore be something like this:

							(Total)
	x	x	x	x	x	x	x
	x	x	x	x	x	x	x
	x	x	x	x	x	x	x
	x	x	x	x	x	x	x
(Artificial break)							
	x	x	x	x	x	x	x
	x	x	x	x	x	x	x
	x	x	x	x	x	x	x
	x	x	x	x	x	x	x
(Total)	x	x	x	x	x	x	x

Such a pattern is not possible if some of the row captions take up more than one line; wherever possible, row captions should be made short enough to fit on one line.

On the question of ruled lines (called 'rules' in printing terms), the basic principle is to use no more than necessary. If proper care is taken with table layout and row and column spacing, no more than five horizontal lines are needed for a two-way table:

	(Coverage)				Units
Line 1–					
	Factor A	Factor B			Total
Line 2–					
		B1	B2	B3	B4
Line 3–					
	A1	x	x	x	x	x
	A2	x	x	x	x	x
	A3	x	x	x	x	x
	A4	x	x	x	x	x
Line 4–					
	Total	x	x	x	x	x
Line 5–					

(Table heading)

(Source)

Each line serves a purpose; further lines would be superfluous. Sometimes the general column headings 'Factor A' and 'Factor B' can be omitted without ambiguity and line 2 is not then required. Given careful control over row spacing in a printed table, lines 3 and 4 can sometimes be omitted without loss of clarity (as, for example, in Table 3.6) but it is generally safer to put them in. Some people prefer lines 3 and 4 to be solid, rather than broken; this is a matter of personal taste. All the example tables in this chapter (except those given as examples of poor layout or style) demonstrate correct use of ruled lines.

Some presenters prefer to avoid using vertical lines ('rules') altogether, while others favour using them either to box in the table, or to separate a 'totals' column from the body of the table, or both. This is another matter of personal taste. However vertical lines should never be drawn between columns in the body of a table: this merely prevents the eye moving smoothly across the rows.

Tables should be printed upright on the page and not sideways: it is irritating to have to twist a book or report in order to read a table (or chart). Occasionally there may be a sufficiently good reason to break this rule for reference tables, but it should never be broken when designing a demonstration table.

Choice of typeface and use of capital letters
The objective here is to adopt a style which is easy to read, pleasing to look at and consistent throughout. The choice of typeface is a personal one but plain kinds are generally best. Within the same publication, all tables (and charts) should conform to the same pattern in their use of bold and italic type and in the size of type used. In practice, the options available will depend on whether the final table is to be printed or typed and, if typed, on what typewriter. However, with care and attention to detail, satisfactory tables can be produced on any kind of typewriter.

The following guidelines are offered as one possible recipe for achieving a consistent and effective style.

Table titles
All tables must have a clear informative title as described in section 3.2, and tables should generally be numbered (see pararaph, Numbering of tables). Where a table is numbered, its title should be above the table, immediately following the table number but separated from it by a space: where a second line is needed it should begin under the first word of the title, not under the table number.

The title should be in lower case letters using capital letters sparingly. If the table title can be produced in bold print or in a larger typeface than that used in the main body of the table, then capitals should be used only for the first letters of the initial word and of proper nouns. If the same size and weight of typeface is used for the title as for row and column headings, then capital letters should normally be used for the initial letters of all but minor words in the title. (Minor words include words like of, by, and, the, etc.)

Row and column headings
Row and column headings should be in the same typeface as the rest of the table, and bold or italic type should be used only when needed to make distinctions, for example to indicate major sub-headings.

In row and column headings only the initial word and proper names should begin with a capital letter.

As far as possible column headings should be horizontal and not sideways on.

Where column headings are of varying length and take up varying numbers of lines all the column headings should begin on the same line and those of fewer lines should not be centred in the vertical space. The exact positioning of column headings is a matter of personal taste and convenience.

If a row description runs to two or more lines, the second and subsequent line(s) should be inset. If a column heading runs to two or more lines, the second and subsequent line(s) should not be inset.

Where a row category is further divided into a number of sub-categories, this should be indicated by in-setting the row headings. See, for example, Table 3.11.

Choice of size and style of typeface for numbers
Statistical tables are usually printed in either six point or eight point type (approximately 12 or 10 characters per inch). The choice is largely a matter of personal preference. Some experiments have been carried out[1] to establish which size is easier to read accurately and the results have always been marginally in favour of the larger typeface. However no difference has been great enough for the experimenter to be sure that the apparent advantage was not due to random variation between the people taking part in the experiment.

It is important that all the numbers should be legible and that the table should look neat. Some people find that a table typed using 12 characters to the inch (the normal

[1] TINKER, M A. Legibility of mathematical tables. *Journal of Applied Psychology*, 1960, vol 44 no 2, pp 83–87.

Table 3.11 Persons received into custody on remand: by sex, age, verdict, and sentence, 1980

England & Wales   Percentages and numbers ☑

	Males				Females			
	14–16	17–20	21 and over	All ages	14–16	17–20	21 and over	All ages
Found not guilty or not proceeded with	1	2	4	3	–	2	5	4
Found guilty								
Given a non-custodial sentence								
Absolute discharge	–	–	1	1	2	1	2	1
Conditional discharge	4	2	3	3	7	6	8	8
Probation order	–	8	6	7	3	22	14	17
Supervision order	4	–	–	–	10	–	–	–
Fine	3	6	8	7	1	7	8	8
Community service order	–	8	3	4	0	4	1	2
Care order	4	0	–	–	10	0	–	–
Suspended sentence	–	5	11	8	0	7	12	10
Other	9	4	5	5	8	8	10	9
Total given a non-custodial sentence	24	33	37	35	41	55	55	55
Given an immediate custodial sentence	71	56	47	52	51	27	26	27
Verdict/sentence not known	4	9	12	10	8	16	14	14
Total remanded in custody (= 100%) (numbers)	3,265	17,198	29,878	50,341	100	1,332	2,264	3,696

Source: Home Office

typeface spacing) looks neater if it is photo-reduced by a factor of 1.4. (This is the reduction necessary to reduce A3 to A4.) Tables 3.12, 3.13 and 3.14 show the same table represented using 10 characters to the inch, 12 characters to the inch, and the 12 characters to the inch reduced by a factor of 1.4. The most suitable size for any particular display is best decided on by such experimentation.

If the table is to be printed in the final version rather than typed, then footnotes will normally be printed using a smaller type size. It is important to ensure that, whatever size of typeface is chosen for figures, any numbers which appear in footnotes should remain legible.

The use of bold type for totals and sub-totals was discussed in section 3.2. It is sometimes useful to employ italics for percentages and rates when these appear in the same table as actual numbers (although the distinction between plain and italic numbers is not always clear). In Table 5.4 (on page 53) for example, the standard errors are shown in italics. If, as was recommended in section 3.2, percentages and rates are assembled together in a separate part of the table, the need for a different typeface is less pressing.

Numbering of tables
Table numbers may be consecutive throughout a publication or alternative methods may be used. Where con-

venient, separate sets of tables may be numbered in sections. For example, tables may be numbered first by the chapter number and then by their position within the chapter (as in this book). Clearly the same number must not be used twice in the same publication.

Tables, graphs and figures are generally given separate series of numbers. The series of numbers may be reduced to two by having one series for tables and another for figures which will include graphs, maps, charts and pictures.

It may not be necessary to number every demonstration table. If a table of data appears in the middle of its supporting text *and* if it is not referred to at any other point in the report then a table number is not strictly necessary. In practice, however, it is often desirable to refer back to a data display and this can be done easily if every data display is numbered. If you decide not to number a particular table and, later, reverse this decision in order to refer back to it, the process of renumbering subsequent tables and adjusting references to them is time-consuming and prone to error.

Punctuation marks for large numbers
Punctuation marks of some kind are essential in large numbers, to mark the position of digits representing thousands and millions. (1,270,000 should never be entered in a table as 1270000.) Although a comma is

Table 3.12 (Table 3.7 repeated: 10 characters per inch)
Marriages by manner of solemnisation

Great Britain

	1966	1971	1976	1977	1978
Manner of Solemnisation					Thousands
Church of England/Church in Wales	175	160	120	117	120
Church of Scotland	22	20	16	15	15
Roman Catholic Church	51	48	34	34)	
Other Christian	38	37	31	32)	70
Jews and other non-Christian	2	2	1	1)	
Register Office	138	180	193	195	200
Total Marriages	426	447	396	394	405
Marriages solemnised in Register Offices as percentages of all marriages					Percentages
England and Wales	33	41	50	51	51
Scotland	25	31	38	38	40
Great Britain	32	40	49	49	49

Source: OPCS; Register Office for Scotland

Table 3.13 (Table 3.7 repeated: 12 characters per inch)

Great Britain

	1966	1971	1976	1977	1978
Manner of Solemnisation					Thousands
Church of England/Church in Wales	175	160	120	117	120
Church of Scotland	22	20	16	15	15
Roman Catholic Church	51	48	34	34)	
Other Christian	38	37	31	32)	70
Jews and other non-Christian	2	2	1	1)	
Register Office	138	180	193	195	200
Total Marriages	426	447	396	394	405
Marriages solemnised in Register Offices as percentages of all marriages					Percentages
England and Wales	33	41	50	51	51
Scotland	25	31	38	38	40
Great Britain	32	40	49	49	49

Source: OPCS; Register Office for Scotland

Table 3.14 (Table 3.13 reduced by 1.4)

Great Britain   ☑

	1966	1971	1976	1977	1978
Manner of Solemnisation				Thousands	
Church of England/Church in Wales	175	160	120	117	120
Church of Scotland	22	20	16	15	15
Roman Catholic Church	51	48	34	34)	
Other Christian	38	37	31	32)	70
Jews and Other non-Christian	2	2	1	1)	
Register Office	138	180	193	195	200
Total Marriages	426	447	396	394	405
Marriages solemnised in Register Offices as percentages of all marriages				Percentages	
England and Wales	33	41	50	51	51
Scotland	25	31	38	38	40
Great Britain	32	40	49	49	49

Source: OPCS; Register Office for Scotland

customarily used in the UK for this purpose, and should always be used in typed tables, a better method for printed tables is to ask the printer to omit the comma itself but space the digits as though the comma were there. The resulting small gaps will be quite sufficient to separate the thousands from the hundreds, and the absence of commas will produce a clearer table.

Footnotes
Footnotes are a necessary evil. Inches of footnotes after a modest table may produce an unbalanced effect and are likely to be ignored; on the other hand it is essential that all the information necessary for the correct inter-pretation of the table should be included in the table design. Footnotes will often therefore be necessary to:

—give fuller definition of a brief heading

—explain any changes in definition which have affected the data in the table

—explain any apparent anomalies or inconsistencies in the table.

Footnotes should be listed one under the other and placed at the extreme left below the table. The style of type for the footnotes should be the same as for the table but smaller if that is convenient and the size is still above the minimum for readability.

Footnotes should be referred to by superior arabic numbers. Letters in brackets or parentheses may be used as an alternative when needed to avoid confusion. Special symbols, such as asterisks and daggers may also be used. However these symbols are not suitable when more than two or three footnotes are required, and are not always available on word processors or standard typewriters.

The indicator of the footnote should be above and to the right of the relevant entry, and the type size for the reference symbol should generally be smaller than for figures in the tables.

Footnotes should be referenced across the columns starting from the lefthand side and the top row—which may be the title—and proceeding similarly for each successive row.

The footnote 'Components may not add to totals (or the total) because they have been rounded independently' should be marked against the appropriate headings.

The use of many of these guidelines is illustrated in Table 3.15, taken from *Social Trends* 10.

In this table the title indicates clearly that the table shows the socio-economic group of people in employ-ment (why should official tables refer to 'persons' rather than people?) sub-divided by ethnic group and sex. The data refer to England in 1977; the entries are (mainly) per-centages and the source of information was the *National Dwelling and Housing Survey* carried out for the Depart-ment of the Environment.

There are two major sections of the table, males and females, with these sub-headings clearly shown in bold typefaces and the socio-economic group labels are inset at two different levels below. Socio-economic group is first

Table 3.15 Socio-economic group of persons in employment; by ethnic group and sex, 1977

England   *Percentages*

	White	West Indian	African	Indian, Pakistani, Bangladeshi	Other[1]	Total
Males: socio-economic group						
Non-manual:						
Professional, employers, and managers	*23*	*4*	*12*	*16*	*27*	*23*
Other non-manual	*18*	*7*	*28*	*11*	*19*	*18*
All non-manual	*41*	*11*	*41*	*26*	*46*	*40*
Manual:						
Skilled manual	*40*	*49*	*41*	*33*	*28*	*39*
Semi-skilled manual	*12*	*25*	*14*	*27*	*18*	*13*
Unskilled manual and other[2]	*8*	*15*	*5*	*13*	*8*	*8*
All manual	*59*	*89*	*59*	*74*	*54*	*60*
Total sample size (=100%)	*100*	*100*	*100*	*100*	*100*	*100*
(numbers)	50,924	519	74	922	757	53,196
Females: socio-economic group						
Non-manual:						
Professional, employers, and managers	*6*	*1*	*3*	*5*	*8*	*6*
Other non-manual	*53*	*45*	*54*	*37*	*52*	*53*
All non-manual	*59*	*47*	*56*	*42*	*61*	*59*
Manual:						
Skilled manual	*8*	*4*	*5*	*12*	*6*	*8*
Semi-skilled manual	*24*	*39*	*31*	*39*	*26*	*24*
Unskilled manual and other[2]	*9*	*11*	*8*	*7*	*8*	*9*
All manual	*41*	*53*	*44*	*58*	*39*	*41*
Total sample size (=100%)	*100*	*100*	*100*	*100*	*100*	*100*
(numbers)	32,502	460	39	309	484	33,794

[1] Includes Chinese, Other Asian, Arab, Other, and Mixed Origin.
[2] Includes farm workers and members of the armed forces.

Source: National Dwelling and Housing Survey, *Department of the Environment*

categorised as 'non-manual' or 'manual' and then further sub-divided into 'professional, employers, and managers' then 'other non-manual' in the first case and into 'skilled manual', 'semi-skilled manual' and 'unskilled manual and others' in the second.

Most of the entries in the table are percentages and these figures are in italic type, as are the corresponding row headings. The exceptions to this are the two rows giving the total sample sizes for males and females where actual numbers are given in non-italic or roman typeface.

The sub-totals 'all non-manual' and 'all manual' are separated from other entries in the column by additional, helpful horizontal lines in the body of the table and by blank spaces in the row headings column. These guide the eye effectively to the two sub-totals which together add to the total of 100 per cent in each section of the table.

The columns are regularly spaced (despite different lengths of column headings) and as close together as can

be achieved with the long column heading which includes 'Bangladeshi'.

A vertical line has been used to separate the 'total' column from the rest: this could probably have been dispensed with without loss of clarity since there is a bigger gap between the last two columns than between the columns in the main body of the table.

The four numbers over 1,000 include commas rather than small spaces but, as there are only four of them the effect is minimal.

Footnotes have been used to amplify the category 'other' as it is applied to ethnic groups and to socio-economic groups. The footnotes are in the same typeface as the rest of the table, but smaller.

It is interesting to see how this same table could be laid out using a standard typewriter and following strictly the recommended guidelines. This has been done in Table 3.16.

Table 3.16 Men and women in employment: by ethnic group and socio-economic group

England 1977   Percentages and numbers

	White	West Indian	African	Indian, Pakistani, Bgldeshi.	Other[1]	Total
Males						
Socio-economic group						
Non-manual:						
Professional, etc[2]	23	4	12	16	27	23
Other non-manual	18	7	28	11	19	18
All non-manual	41	11	41	26	46	40
Manual:						
Skilled	40	49	41	33	28	39
Semi-skilled	12	25	14	27	18	13
Unskilled, etc	8	15	5	13	8	8
All manual	59	89	59	74	54	60
Total sample size = 100% (Numbers)	50,924	519	74	922	757	53,196
Females						
Socio-economic group						
Non-manual						
Professional, etc[2]	6	1	3	5	8	6
Other non-manual	53	45	54	37	52	53
All non-manual	59	47	56	42	61	59
Manual:						
Skilled manual	8	4	5	12	6	8
Semi-skilled manual	24	39	31	39	26	24
Unskilled and other[3]	9	11	8	7	8	9
All manual	41	53	44	58	39	41
Total sample size = 100% (Numbers)	32,502	460	39	309	484	33,794

[1] Includes Chinese, Other Asian, Arab, Other, and Mixed Origin
[2] Professional, employers and managers
[3] Includes farm workers and members of the Armed Forces

Source: National Dwelling and Housing Survey Department of the Environment

Using a standard typewriter the two main problems are the long row and column headings and the 'non-manual' and 'manual' sub-totals.

In order to fit the table upright onto an A4 page, leaving equal amounts of space between all columns except the last two (in order to isolate the 'total' column from the rest of the table) it is necessary to rearrange the sub-headings 'Males: socio-economic group' and 'Females: socio-economic group'. The long column heading 'Bangladeshi' has been abbreviated for the same reason.

The sub-totals have been offset, as recommended in section 3.5 (to the right rather than to the left after experimenting with both) and, with the use of ruled lines to isolate them from the other numbers in the same column, their different status and the fact that the 'non-manual' plus 'manual' sub-totals add to 100 per cent is fairly clear.

Without an italic typeface to differentiate percentages from numbers, the table looks cluttered if the 100% total is repeated under each column and so this row has been omitted. Similarly, since no bold typeface is available for the main table heading or sub-headings, these have all been underlined to give them greater emphasis.

The final table does not have the neat appearance of the printed version in *Social Trends* but it is clear and easy to interpret.

It is reproduced, reduced by an A4 to A5 reduction factor, below.

Table 3.17 Men and women in employment: by ethnic group and socio-economic group

England 1977 — Percentages and numbers

	White	West Indian	African	Indian, Pakistani, Bgldeshi.	Other[1]	Total
Males						
Socio-economic group						
Non-manual:						
Professional, etc[2]	23	4	12	16	27	23
Other non-manual	18	7	28	11	19	18
All non-manual	41	11	41	26	46	40
Manual:						
Skilled	40	49	41	33	28	39
Semi-skilled	12	25	14	27	18	13
Unskilled, etc	8	15	5	13	8	8
All manual	59	89	59	74	54	60
Total sample size = 100% (Numbers)	50,924	519	74	922	757	53,196
Females						
Socio-economic group						
Non-manual						
Professional, etc[2]	6	1	3	5	8	6
Other non-manual	53	45	54	37	52	53
All non-manual	59	47	56	42	61	59
Manual:						
Skilled manual	8	4	5	12	6	8
Semi-skilled manual	24	39	31	39	26	24
Unskilled and other[3]	9	11	8	7	8	9
All manual	41	53	44	58	39	41
Total sample size = 100% (Numbers)	32,502	460	39	309	484	33,794

[1] Includes Chinese, Other Asian, Arab, Other, and Mixed Origin
[2] Professional, employers and managers
[3] Includes farm workers and members of the Armed Forces

Source: National Dwelling and Housing Survey Department of the Environment

Chapter 4: Demonstration tables

4.1 Which numbers to present

In demonstration tables the prime function of the table is to *communicate a message*. This means that you, the writer, must first decide what message to present by careful analysis of the data, and then design a table to illustrate that message as effectively and memorably as possible.

Most data can be presented in a number of different ways: for example, each number can be expressed as a percentage of the appropriate row or column total; data recorded over time can be expressed as a series of index numbers based on a specific year whose value is taken as 100; the ratio or difference between two rows or columns may be calculated; and incidence of accidents or disease may be quoted as rates per 1,000 at risk; sums of money can be recorded either in current terms or can be measured at constant prices by dividing each number by an appropriate factor; and, of course, the original data can be displayed, usually rounded to two or three figures.

The choice of which numbers to use in a demonstration table depends on the message being illustrated. In practice the process of analysing the original data is likely to involve calculating a number of derived statistics (differences, ratios, percentages and averages) and, by the time the main patterns in the data have been identified, the most appropriate way of highlighting these patterns is likely to be clear to the analyst.

At the exploratory stage some patterns in the data may be revealed by calculating fairly complicated derived statistics (for example, ratios of successive differences or percentages of percentages) but in presenting the data it is important to avoid using over-elaborate derived statistics.

Few readers will have any difficulty in interpreting a table of percentages (so long as it is clear which total the percentages are based on), or a table of numbers quoted as indices based on a specific year (for example, 1975 = 100) but many will quail before a column headed 'ratio between male and female upper quartiles'.

Section 4.2 lists a number of derived statistics and gives examples of how they can be used effectively. The list is not exhaustive and other derived measures may be found which illustrate particular messages effectively. If in doubt about the effectiveness of a possible table of derived statistics, ask yourself 'will my mother understand this table?'.

4.2 Some derived statistics

Percentages

Percentages are useful when changes in the composition of a total are of particular interest. For example, Table 4.1 (a copy of Table 1.1) shows that, in Wales, a considerably smaller percentage of entrants to the Youth Opportunities Programme joined Work Experience on Employers' Premises (WEEP) schemes in 1979/80 than in 1978/79.

Since the total number of entrants to the programme changed considerably over the period (from 15,000 in 1978/79 to 22,000 in 1979/80), if the absolute numbers of entrants to each scheme had been quoted rather than the percentages, the change in overall pattern would have been much less obvious.

Table 4.1 Entrants to Youth Opportunities Programme in Wales: by type of scheme

Wales 1978 to 1980   *Percentages*

	1978/79	1979/80
WEEP[1]	89	73
Short Training Course	10	10
Community Service		9
Project based work experience		7
Training workshops	1	1
Induction and other[2]		1
Total (100%)	15,000	22,000

[1] Work Experience on Employers' Premises
[2] Employment induction courses and other remedial and preparatory courses

When a demonstration table shows percentages the totals on which the percentages are based should always be quoted if they are not given elsewhere. There are two reasons for this: first, quoting the totals allows an interested reader to explore the table more thoroughly by working out the numbers in each category (for example, the number of young people on WEEP schemes actually increased between 1978/79 and 1979/80 from about 13,400 to over 16,000, although this scheme accounted for a considerably smaller percentage of the total in 1979/80 than in 1978/79); and secondly, it provides necessary background information against which the significance of changes in percentages can be assessed (a change from 30 per cent to 33 per cent on a total of 100 represents only

Table 4.2 School leavers by A-level specialisation and destination

England 1978 to 1979

	Percentage going to			Total numbers with two or more A-levels
	Degree courses	Other further education	Employment	
Leavers with two or more A levels in:	%	%	%	(000s =100%)
Science inc maths	76	5	20	20.8
Science without maths	63	11	25	9.6
Combination inc science	56	14	30	21.0
Arts	50	20	30	18.7
Social science	32	20	47	3.2
Arts/social science	51	17	33	18.8
All subjects	58	14	28	92.2

Source: Statistics of Education 1979 Vol 2,
Department of Education and Science

three individuals: the same change in percentages based on a total of 10,000 represents 300 individuals and is clearly a more significant change). Naturally, if the original numbers are of fundamental importance in their own right, they should be displayed as well as the percentages. When both the numbers and percentages are displayed, it is generally best to put them in separate tables, or separate parts of the same table.

Tables may contain percentages of either column totals, as in Table 4.1, or row totals, as in Table 4.2.

In both cases, the table should be accompanied by a verbal summary highlighting the main pattern. The verbal summary which accompanied Table 4.1 was quoted in Chapter 1 (page 12). The original article from which Table 4.2 is taken included the following commentary:

> . . . '[Table 4.2] gives an indication of how the subject choice at A level is linked with the pupils' destination on leaving school. It shows that leavers with two or more passes, at least one of which is a science subject are more likely to go on to degree courses than those with passes in arts or social science subjects only. The latter are more likely to go on to other types of full-time education or to seek employment on leaving school . . .'

Indices
When the reader is invited to compare the growth of two or more measurements over a number of years, differences in growth patterns may be hard to detect if each series has a different starting value. For example Table 4.3 shows sales of gas to different types of consumer from 1975 to 1978.

Table 4.3 Gas sales[1] by type of consumer

Great Britain — *Million therms*

	Domestic	Industrial	Commercial	Public Admin[2]	Total
1975	5 870	5 870	1 170	170	13 080
1976	6 170	6 260	1 340	190	13 970
1977	6 570	6 400	1 370	210	14 550
1978	7 240	6 300	1 510	230	15 380

[1] Public supply system
[2] National and local government, including public lighting

Source: Department of Energy

It is clear from the table that sales to each type of consumer have increased over the period, but it is not clear which category of sales has shown the greatest percentage growth. In Table 4.4 the sales in each year have been expressed as a percentage of their 1975 level: thus the 1976 value for domestic gas sales is given as 105 (that is $\frac{6170}{5870} \times 100$) and the table is labelled to indicate that the entries are indices based on 1975 as 100.

Table 4.4 Gas sales[1] by type of consumer

Great Britain — *Index: 1975 = 100*

	Domestic	Industrial	Commercial	Public Admin[2]	Total
1975	100	100	100	100	100
1976	105	107	115	113	107
1977	112	109	117	123	111
1978	123	107	129	135	118

[1] Public supply system
[2] National and local government, including public lighting

Source: Department of Energy

Table 4.5 Shops, number of employees and turnover by type of retail organisation

Great Britain 1961 to 1971

	Numbers of shops (000's)			Number of employees (000's)			Turnover (£m)		
	Co-ops	Mults[1]	Indeps	Co-ops	Mults	Indeps	Co-ops	Mults	Indeps
1961	29	67	446	200	630	1 160	960	2 580	5 290
1966	27	74	404	170	740	1 640	1 020	3 840	6 280
1971	15	67	403	130	820	1 610	1 100	6 060	8 060

[1] Multiples refer to shops belonging to retail organisations with 10 or more branches

Source: Provisional results of the Census of Distribution 1971. *Department of Trade and Industry*

From Table 4.4 it is immediately clear that total sales of gas increased by 18 per cent between 1975 and 1978, that the greatest percentage increase was in gas sold to national and local government and that gas sold to commercial users increased by 29 per cent.

Ratios
Table 4.5 gives the number of shops, the number of people employed and the total turnover in three different types of retail organisations: co-operatives, multiples and independents, in 1961, 1966 and 1971.

A first examination of the table reveals that over the period 1961 to 1971 the number of co-operative and independent retail shops declined, the number of people employed by these organisations also declined while the annual turnover increased. The number of people employed in multiples increased, the turnover increased and the number of shops in 1971 was the same as the number in 1961. The alert reader is immediately interested in deriving a number of ratios and examining their patterns: obvious ratios are turnover per shop, turnover per employer and, possibly, employees per shop. Tables 4.6 and 4.7 show the first two of these ratios.

Table 4.6 Average turnover per shop: by type of retail organisation

Great Britain 1961 to 1971 £000's

	Co-ops	Mults	Indeps
1961	33	39	12
1966	38	52	16
1971	73	90	20

From Table 4.6 we see that for both the co-operatives and the multiples, turnover per shop more than doubled between 1961 and 1971 (from £33,000 to £73,000 in the first case, and from £39,000 to £90,000 in the second) while, for the independents turnover per shop increased by 66 per cent, from £12,000 to £20,000.

A similar pattern emerges from Table 4.7: the average turnover per employee in co-operatives and multiples

increased by about 80 per cent between 1961 and 1971 while, for the independents, the corresponding increase was 56 per cent (from £3,200 to £5,000).

Table 4.7 Average turnover per employee: by type of retail organisation

Great Britain 1961 to 1971 £000's

	Co-ops	Mults	Indeps
1961	4.8	4.1	3.2
1966	6.0	5.2	3.8
1971	8.5	7.4	5.0

In cases like this, where two or three related measurements vary over time or between different locations, the ratio between measurements may reveal more informative patterns than the original data.

Differences
In some cases the difference between two rows or columns may be of particular interest. Consider, for example, the four columns in Table 4.8. These figures have been extracted from a more comprehensive table on population change from 1971 to 1981, and here, as in the original table, the 'Natural increase' column appears logically as the difference between the entries in the births

Table 4.8 Natural population change 1971 to 1981

thousands

	Popn at beginning	Births	Deaths	Natural increase
1971–72	48,850	749	580	+169
1972–73	49,030	702	590	+110
1973–74	49,150	652	583	+ 70
1974–75	49,160	625	589	+ 36
1975–76	49,160	594	599	− 5
1976–77	49,140	568	580	− 12
1977–78	49,120	576	584	− 8
1978–79	49,120	626	591	+ 35
1979–80	49,170	647	577	+ 70
1980–81	49,240	645	577	+ 69

Source: Population Trends 30, Table 1

column and the deaths column for each year. This enables the reader to see immediately that between 1975 and 1978 there were more deaths than births each year and to observe the general pattern in natural population change (that is change in the population excluding immigration and emigration) over the ten-year period.

Rates per number at risk
It is important that numbers appearing in the same table should be directly comparable. This may involve re-expressing the original numbers as incidence rates per thousand at risk or per hour of exposure to risk. For example, quoting only the original data as in the first three columns of Table 4.9, might suggest that the RAF is positively careless with its aircraft by comparison with either the Royal Navy or the Army. However, when the number of accidents in each of the services is re-expressed as the number of accidents per 10,000 flying hours, a quite different picture emerges. Measured in these units the RAF's safety record looks altogether more creditable.

Table 4.9 Aircraft accidents involving loss or serious damage: by Service ☑

	Number of Accidents			Rates per 10,000 flying hours		
	Royal Navy	Army	Royal Air Force	Royal Navy	Army	Royal Air Force
1977	8	4	16	0.88	0.39	0.34
1978	8	7	24	0.89	0.69	0.51
1979	3	11	25	0.36	1.18	0.52
1980	5	9	23	0.56	1.01	0.48
1981	8	1	23	0.86	0.11	0.50

Source: Statement on the Defence Estimates 1983,2 Cmnd. 8951-II, Table 6.6

Many accident statistics are more meaningful when quoted as accidents per so many hours exposure to risk (for example accidents to different sorts of sportsmen and accidents associated with different modes of transport). Similarly, absolute numbers of births or deaths are often insufficient to make valid comparisons between regions containing markedly different numbers of women of child-bearing age and old people.

Money measured at constant prices
Particularly in times of high or varying inflation, it may be difficult to discern true patterns of growth or decline in a series of values recorded in money terms. For example, if retail prices have approximately doubled over the last ten years, a shopkeeper whose receipts have risen by only 80 per cent over that period is probably less successful now than he was ten years ago. To reveal true trends in such monetary measurements, the figures should be adjusted to 'constant prices'. This is done by dividing each year's observed value by a deflating factor which takes into account the almost automatic change in the measurement attributable only to inflation. For many series the Retail Price Index will be an appropriate deflating factor; in other cases more specialised index numbers will be used to correct for changes of price in specific categories of goods.

Table 4.10 below shows agricultural output in four categories measured at current prices from 1970 to 1978 while Table 4.11 shows the same data revalued at constant, 1975, prices.

Table 4.10 Agricultural output at current prices ☑

Great Britain   *£ million*

	Livestock	Livestock products	Farm crops	Horti-culture
1970	870	700	460	270
1972	1 070	810	470	330
1974	1 520	1 170	890	470
1976	2 160	1 670	1 450	590
1978	2 780	2 010	1 530	710

Source: Annual Abstract of Statistics 1980 Edition. Table 9.1

Table 4.11 Agricultural output at constant (1975) prices ☑

Great Britain   *£ million*

	Livestock	Livestock products	Farm crops	Horti-culture
1970	1 750	1 320	1 060	550
1972	1 770	1 410	1 060	560
1974	1 890	1 380	1 090	580
1976	1 820	1 420	1 230	510
1978	1 850	1 540	1 240	590

Source: Annual Abstract of Statistics 1980 Edition. Table 9.2

From Table 4.10 it appears that the value of agricultural output in 1978 was approximately three times its 1970 value in the four categories quoted. However, when the data are measured at constant prices, as in Table 4.11, it is clear that the increase in value between 1970 and 1978 was of the order of 10 per cent and that in livestock and horticulture output was lower in 1976 than in 1974.

4.3 Presentation of demonstration tables
Analysis of any set of data is likely to involve producing several tables of derived statistics and roughly sketched graphs. If each such table or graph is accompanied by a brief note of the main patterns revealed by this form of presentation, these notes can be used to decide on the final presentation of the data. Generally, only the most important patterns need to be reported and these are usually (though certainly not always) the most striking ones. The derived tables or graphs which best illustrate these patterns should be included in the final report. If tables are used, the numbers in each table must be presented so as to highlight the main patterns in the data.

Demonstration tables should appear in the main body of text (not as annexes or afterthoughts) where they can amplify the argument and encourage the reader to pause momentarily and consider the numbers. There is no place in the main text of a report for 'some figures which may be relevant'. If you do not know precisely what contribution a table makes to the development of an argument, leave it out: the reader is unlikely to find any pattern in a table unless you direct his or her attention to it. You may of course include reference tables separately, probably at the back of the report.

The general rules on structure and layout, described in Chapter 3, apply to all tables including demonstration tables. Briefly these are:

—The table should contain a clear and complete explanation of what the figures represent (kind of objects enumerated and how they have been categorised in the table; the geographical and time coverage; units of measurement and source of the data);

—Spacing and horizontal ruled lines should be used to guide the eye, with minimum use of vertical ruled lines;

—Use a clear typeface with capital letters only for initial letters of first words and proper nouns;

—Footnotes should be used conservatively but included wherever necessary to prevent misunderstanding or misinterpretation of the data and to give fuller description of compact headings.

Thereafter, all the guidelines recommended for constructing effective demonstration tables are designed to highlight patterns and exceptions in the table. These guidelines were formally stated by Ehrenberg[1] who gives the following criterion for a good table:

'The patterns and exceptions in a table should be obvious at a glance once the reader has been told what they are.'

It is important to note the last few words of this directive: 'once the reader has been told what they are'. Figures very rarely speak for themselves, and even experienced analysts sometimes pause doubtfully when confronted with a table and an injunction to 'just look at these figures!'. When writing for a non-specialist audience it is particularly important that tables (and charts) are used to amplify the text in an obvious and helpful way. The introductory phrase 'As can be seen from Table X.Y . . .' is likely to appear frequently.

The guidelines proposed by Ehrenberg help both to uncover the basic patterns in a set of data and to communicate selected patterns and exceptions to future readers.

4.4 Seven basic rules
These seven basic guidelines are:

1. Round all numbers to two effective digits

2. Put the numbers to be most often compared with each other in columns rather than rows

3. Arrange columns and rows in some natural order or in size order

3a. Where possible, put big numbers at the top of tables

4. Give row and column averages or totals[2] as a focus

5. Use layout to guide the eye

6. Give a verbal summary of the main points of the table

7. Use charts to show relationships.

Each guideline will now be considered in more detail, and the effect of implementing it will be demonstrated using Table 4.12 as a simple illustration.

Rule 1: Round to two effective digits
Effective digits are defined as those which vary within a set of data. If the data consist of three digit numbers and all the digits vary, then the rule will be equivalent to rounding to two significant digits. For example, the following data

| 178 | 242 | 575 | 582 | 633 | 750 |

would be rounded to

| 180 | 240 | 580 | 580 | 630 | 750 |

If, however, the hundreds digit remained constant throughout then all three digits would be retained. Thus the set of numbers

| 112 | 126 | 132 | 148 | 165 | 174 |

would remain unchanged. This is because, given the range of variation in the data (namely from 112 to 174), the two digits which discriminate between one number and another are the tens and units digits: these are the *effective* digits in this set of data.

Similarly, with four digit numbers the rule might dictate that we retain two, three or four digits depending on how many of the digits change within the set of data.

So   1,693 2,832 3,801 4,886 6,181 7,309

would be rounded to

1,700 2,800 3,800 4,900 6,200 7,300

the set

2,114 2,186 2,318 2,468 2,663

would be rounded to

2,110 2,190 2,320 2,470 2,660

and the data

2,211 2,236 2,259 2,274 2,286

would stay unchanged.

[1] EHRENBERG, A S C. Rudiments of numeracy (with discussion), *Journal of the Royal Statistical Society, Series A: General*. 1977, vol 140 para 3, pp 277–297.
[2] Ehrenberg proposed averages only.

Table 4.12 Motor vehicles currently licensed

United Kingdom   *Thousands*

Type of vehicle	1961	1966	1971	1976	1981[5]
Private cars and private vans	6,114	9,747	12,361	14,373	15,632
Motorcycles, scooters, and mopeds[1]	1,842	1,430	1,033	1,235	1,386
Public transport vehicles[2]	94	96	108	115	112
Goods[3]	1,490	1,611	1,660	1,796	1,771
Agricultural tractors, etc	481	478	450	414	373
Other vehicles[4]	206	260	247	300	510[6]
All vehicles	10,227	13,621	15,859	18,233	19,784

[1] Also includes Northern Ireland figures for three wheelers (under 406 kilos) and pedestrian controlled vehicles.
[2] Includes taxis. Tram cars are included in 1961 and 1966.
[3] Includes agricultural vans and lorries, tower wagons, showmen's goods vehicles and vehicles licensed to draw trailers. Also includes Northern Ireland figures for general haulage tractors.
[4] Excludes Northern Ireland. Includes three wheelers, pedestrian controlled vehicles and general haulage contractors.
[5] From 1978 the census has been taken on 31 December, previous censuses were taken on 30 June.
[6] Includes exempt tax classes 61 and 62 (162 thousand vehicles in 1980, 170 thousand in 1981) which were not included in previous censuses.

Source: Department of Transport

The advantage of rounding to two effective digits is that it makes it easy to compare numbers 'at a glance'. Seeing a pattern in a set of data depends upon observing differences or ratios between pairs of numbers: this depends on the reader's ability to do mental arithmetic. Most people can subtract 1,700 from 2,800 mentally, and remember the answer, 1,100, long enough to compare it with the difference between 3,800 and 2,800. Few people, however, can carry out mental arithmetic on three digit numbers. *Without writing anything down*, try to work out whether the difference between 4,886 and 6,181 is bigger or smaller than the difference between 6,181 and 7,309. Now do it with the rounded numbers.

Digits which do not change in a set of data can be mentally filtered out of the calculation, leaving only two digits to be dealt with. Thus, in the second example quoted above, the 2 in the thousands column can be ignored in comparing successive numbers. Using the rounded row,

2,110   2,190   2,320   2,470   2,660

it is easy to verify that the difference between successive pairs of numbers increases as we work across the data. This is much less easy to check using the unrounded numbers

2,114   2,186   2,318   2,468   2,663

The effect of rounding all the numbers in Table 4.12 to two effective digits is as shown in Table 4.13.

The rounding is carried out line by line: first we round the private cars and vans line, then the motorcycles, scooters and mopeds line and so on. This is because any patterns in this data will relate to growth or decline in different types of vehicles and so the numbers which will be compared with each other most often are numbers in the same line of Table 4.13.

The first and third lines in this table illustrate a commonly encountered problem: the number of digits changes—from four to five (6,100 to 15,600) in the case of the first line and from two to three (94 to 112) in the third

Table 4.13 Motor vehicles currently licensed: figures rounded to two effective digits (footnotes omitted)

United Kingdom   *Thousands*

Type of vehicle	1961	1966	1971	1976	1981
Private cars and private vans	6,100	9,700	12,400	14,400	15,600
Motorcycles, scooters, and mopeds	1,840	1,430	1,030	1,240	1,390
Public transport vehicles	94	96	108	115	112
Goods	1,490	1,610	1,660	1,800	1,770
Agricultural tractors, etc	480	480	450	410	370
Other vehicles	210	260	250	300	510
All vehicles	10,200	13,600	15,900	18,200	19,800

line. No hard and fast rule is offered here and a common sense approach must be adopted. In the table above, after the number of private cars and vans licensed passes 10,000 it remains under 20,000 and so, had we wanted to round the last three entries alone to two effective digits, the 1 of 10,000 would not have counted as an effective digit and three non-zero digits would have been retained. So, in this case the numbers are all rounded to the nearest hundred. If, having passed the 10,000 mark, the number of private cars and vans had climbed over 20,000 or 30,000 it would probably have been more helpful to give all the numbers in that line to the nearest thousand, leaving the first entry as 6,000 and rounding the second to 10,000. Similar reasoning has been applied to the entries in the third row of the table, retaining all three digits after the number of public transport vehicles passes 100. All other rows of the table have been simply rounded to two effective digits.

One snag about rounding to two effective digits is that the figures in the table will not necessarily sum to the printed totals: this can be disconcerting to some readers and, except when writing specifically for an audience whom you know to be knowledgeable about such things, it is wise to put in an explanatory footnote, even though this itself detracts from the directness and simplicity of the presentation. (Something like 'Components may not add to totals because they have been rounded independently' is probably all that is required.) *Do not, on any account, change the figures to make them sum to the totals.*

Having rounded the entries in each row to two effective digits, it is much easier to carry out the mental arithmetic necessary to see the main pattern in the table: from 1961 to 1981 the total number of vehicles licensed almost doubled, from 10,200 thousand to 19,800 thousand; virtually all of the increase was attributable to the growth in the number of private cars and vans licensed (from 6,100 to 15,600 thousands); other categories showed relatively small changes and two categories (motor cycles etc and agricultural tractors) declined over the period.

People are frequently reluctant to round numbers to two effective digits on the grounds that they are 'throwing away accuracy'. Certainly accuracy is being sacrificed: what is gained is the ability to communicate a specific quantitative message and that is the purpose of demonstration tables. People do not find numbers easy to remember—but round two-digit numbers are infinitely more memorable than four- or five-digit numbers.

The magnitude of the error incurred can, of course, always be calculated. If three-digit numbers are rounded to two effective digits, the largest percentage error occurs at the lower end of the range, for example when 104 is rounded down to 100 or 105 is rounded up to 110 there is an error of approximately 5 per cent in either case. At the other end of the range, if 994 is rounded down to 990, the percentage error is about 0.4 per cent. Whether or not the loss of this degree of accuracy is significant depends on the data in question. Assuming that the ultimate objective of any demonstration table is to influence decisions, the acid test must be 'would a different decision be indicated if the precise, unrounded numbers were displayed?'. If the answer to this question is 'yes', then clearly it would be wrong to round the numbers—but, in such a case, searching questions must be asked about the accuracy of the processes of collection, collation and analysis of the data in order to be completely confident that small differences between comparatively large numbers really exist after allowing for the inevitable errors in these processes.

The other reason that is given for reluctance to round data to two effective digits is that the numbers are then not accurate enough to use in further calculations. This is absolutely true: they are not. However, the numbers in a demonstration table are intended to illustrate patterns and exceptions as effectively as possible; they are not intended for use in further calculations. If you consider it likely that a reader will want to do further calculations based on the numbers in a demonstration table, it is important that he or she should be able to refer to the original data as easily as possible. This can be achieved by giving precise details of the source of the data, if necessary including an address and telephone number for enquiries as a footnote, or, in extreme cases, including a reference table as an appendix to the report. There the reader should find more precisely recorded figures plus an indication of the size of the intrinsic error associated with them, and these are the numbers for use in further calculations.

Rule 2: Put numbers to be compared with each other in columns, not rows
The reasoning behind this rule is very straightforward. Firstly, comparisons will often involve subtracting one number from another: we all learned to do subtraction sums by putting the larger number above the smaller and writing the answer underneath; we were trained to subtract in columns. Secondly, it is much easier for the eye to ignore digits which are common to both numbers when the numbers are vertically aligned. For example, in the case of five-digit numbers where the first digit is common throughout we might have

    18,200   14,700   13,500

or 18,200
    14,700
    13,500

When the numbers are printed across the page more effort is required to verify that the leading one is common to all numbers and that they all end in two zeros. When they are printed downwards these points are obvious at a glance.

So we have to decide which comparisons will be made most frequently. Clearly this depends on the message(s) which the table illustrates and thus no general rule will apply. However, when data have been recorded over a number of years, time trends in individual categories will often be important. Here this would indicate that the table used as our example be redrawn with rows and columns interchangéd, as in Table 4.14.

Table 4.14 Motor vehicles currently licensed: rows and columns interchanged (footnotes omitted)

United Kingdom, 1961 to 1981   *Thousands*

Year	Type of vehicle						All vehicles
	Private cars and vans	M'cycles scooters, mopeds	Public transport	Goods vehicles	Agric. tractors etc	Other vehicles	
1961	6,100	1,840	94	1,490	480	210	10,200
1966	9,700	1,430	96	1,610	480	260	13,600
1971	12,400	1,030	108	1,660	450	250	15,900
1976	14,400	1,240	115	1,800	410	300	18,200
1981	15,600	1,380	112	1,800	370	510	19,800

Source: Department of Transport

When rows and columns are interchanged care must be taken to keep the columns evenly spaced. Problems can arise when column headings are of different length but, in such cases, long column headings should be spread over several lines, if necessary breaking long words (eg agricultural) with a hyphen.

Rules 3 and 3a: Arrange columns and rows in some natural order of size: where possible put big numbers at the top
These rules again relate to highlighting patterns and exceptions within the data. If your table includes data about a number of different towns, a natural order might be in decreasing order of population size; if it relates to countries, one natural order might be in decreasing order of wealth, as measured by GDP per head, and another might be in order of physical area. A number of measurements might reasonably be expected to follow such a 'natural' ordering: for example the number of local schools, or doctors or policemen or supermarkets might be expected to follow the same order as the population of towns; measurements associated with wealth, such as the number of cars or telephones per head of population, or the amount spent on education per head of population might be expected to follow the same order as GDP per head. Thus when measurements in a table break this 'natural' sequence the 'surprising' order is highlighted.

Where there is no 'natural' order rows and/or columns should be arranged as far as possible in decreasing order of row or column averages. In Table 4.14 this will involve rearranging the columns so that goods vehicles appear after private cars and vans and before motorcycles etc (there were more motorcycles etc than goods vehicles in 1961, but in all later years there were more vehicles in the goods category); the fourth category will be agricultural tractors—and then there is a problem because the 'other vehicles' category is larger than the public transport vehicles. When this happens the rule should be ignored: any catch-all category such as 'others', 'miscellaneous' or 'unknown' should appear at the end of the data. Frequently such a category will be the smallest and so the regular pattern of decrease will be preserved.

In many tables the order in one direction (either rows or columns) will be dictated by the need to follow a chrono-logical order. In Table 4.14, clearly the years determine the ordering of the rows.

The guideline which recommends placing big numbers at the top of columns whenever possible is again designed to help the reader carry out subtraction sums easily. It is indeed easier to calculate the difference between 9,700 and 6,100 when they appear like this

9,700
6,100

than when they appear like this 6,100
9,700

This would therefore suggest that the most recent year be placed at the top of the table and earlier years beneath it. For some audiences this may be helpful but not for an audience accustomed to using government statistical tables. In virtually all official statistical publications, time is shown as moving either from left to right or from top to bottom, with the most recent year on the extreme right-hand side or at the bottom of a table. Since the aim of a demonstration table is that the reader should find it easy to interpret, it would be thoroughly counter-productive to present the years in an unexpected order.

Where time is neither a row heading nor a column heading then the guidelines should be followed—with, as far as possible, the biggest numbers at the top lefthand corner of the table and a general pattern of decrease from that point.

So, arranging the columns of Table. 4.14 we have Table 4.15.

Three minor points are worth mentioning before leaving the subject of order of rows and columns.

First, for demonstration tables, alphabetic order is not a 'natural' order. In a reference table it is of course helpful to have rows arranged in alphabetic order, so that the reader can look up a desired entry as efficiently as possible: but a demonstration table should be designed to highlight the patterns in the table—not for ease of reference.

Table 4.15 Motor vehicles currently licensed: columns re-arranged in decreasing order of averages (footnotes omitted)

United Kingdom, 1961 to 1981   *Thousands*

Year	Type of vehicle						All vehicles
	Private cars and vans	Goods vehicles	M'cycles, scooters, mopeds	Agric. tractors etc	Public transport	Other vehicles	
1961	6,100	1,490	1,840	480	94	210	10,200
1966	9,700	1,610	1,430	480	96	260	13,600
1971	12,400	1,660	1,030	450	108	250	15,900
1976	14,400	1,800	1,240	410	115	300	18,200
1981	15,600	1,800	1,380	370	112	510	19,800

Source: Department of Transport

Secondly, if you are not confident that your reader is aware of a particular 'natural' ordering (for example, if he or she does not know the populations of the towns or countries included in the table) you should clarify the point by stating explicitly what criterion has determined the row and/or column order.

Finally, if two or three tables with the same row or column headings appear in a report, the same order should be preserved in all tables. If there is no external measure of size to dictate the row and column orders then all the tables should be considered together in order to choose the order which will be most suitable for the group of tables taken together.

Rule 4: Give row and column averages or totals as a focus
The reason for this rule is that averages (where they are appropriate) allow the reader to see an overall pattern by scanning the margins of a table. Having absorbed the overall pattern, he or she can then examine the body of the table to see whether the general pattern is repeated in each row of the table, and also to see how much variation there is about each average.

In our example table it would make little sense to work out row averages, that is an average for each year of the private cars and vans, the public transport vehicles, the tractors and the other vehicles—this would be quite meaningless. The annual total, which is given, is the figure which is appropriate. It is, however, informative to include an average for each type of vehicle calculated for the five years quoted in the table that is the column averages. These are shown in Table 4.16.

A quick glance at the averages in the bottom row of the table reveals the following points:

—On average, private cars and vans accounted for roughly three-quarters of all vehicles licensed (even more drastic rounding to 12,000/16,000);

—Only the first three categories (private cars and vans, goods, and motorcycles) have averages of over 1,000 thousand: by comparison with the private cars and vans, the remaining categories are very small indeed.

We can now investigate the table year by year to see whether or not these three patterns are true throughout the period. We find:

—it is not until 1971 that private vehicles assume quite such dominance; in 1961 and 1966 they account for 60 to 70 per cent of all vehicles licensed, whereas from 1971 onwards they account for roughly three-quarters of each annual total;

Table 4.16 Motor vehicles currently licensed: column averages included (footnotes omitted)

United Kingdom, 1961 to 1981   *Thousands*

Year	Type of vehicle						All vehicles
	Private cars and vans	Goods vehicles	M'cycles, scooters, mopeds	Agric. tractors etc	Public transport	Other vehicles	
1961	6,100	1,490	1,840	480	94	210	10,200
1966	9,700	1,610	1,430	480	96	260	13,600
1971	12,400	1,660	1,030	450	108	250	15,900
1976	14,400	1,800	1,240	410	115	300	18,200
1981	15,600	1,800	1,380	370	112	510	19,800
Average	11,600	1,670	1,390	440	105	300	15,500

Source: Department of Transport

—in 1961 there were more motorcycles, scooters and mopeds than goods vehicles; thereafter the order of categories is the same as that in the average row;

—in all years it is only the first three categories which have totals over 1,000 thousand.

Finally we can examine individual columns of the table to see how much variation there is about each average and to see whether there is any general relationship between pairs of columns. This reveals:

—considerable variation in the first column: from 6,100 (well below the average) to 15,600 (well above the average). The final entry is roughly two and a half times the first entry;

—comparatively little variation about the average in the other columns (the apparent growth from 210 to 510 in the 'other vehicles' category is attributable to the addition of 170 thousand vehicles which were not included in previous censuses: this is made clear in footnote 6 of the original version of our example table—Table 4.12): except for the third column and the 'other vehicles' column, the average coincides with a value roughly half way through the period;

—a general decline in the proportion of motorcycles, scooters and mopeds: in 1961 there were just over three times as many private cars and vans as motorcycles; in 1981 there were over eight times as many: viewed differently, in 1961 the motorcycle category accounted for just under 20 per cent of all vehicles;

—the motorcycles category is unusual in that its numbers declined sharply from 1961 to 1971 and then started to climb, although not to the 1961 level.

Not all of these observations will appear in the final summary of the table, but using the column averages as a focus has helped to identify a number of patterns and exceptions.

In many tables a choice has to be made between showing row or column averages and showing row or column totals. Averages are sometimes more helpful in interpreting the table than totals. This is because averages are of the same order of magnitude as the entries in the table and can therefore be used as a focus when investigating the patterns of variations within a row or column: by contrast the total of a row with five entries is roughly five times as big as the individual entries.

Rule 5: Use layout to guide the eye
This rule is included here only because it is one of the seven basic rules formulated by Ehrenberg (1977). The recommendations are exactly the same as those discussed in Chapter 3 under the heading 'Spacing of rows and columns and use of ruled lines'. Briefly they are:

—Rows and columns within the body of the table should be regularly spaced;

—Additional blank spaces should be used to separate a 'total' or 'average' row or column from the body of the table;

—Horizontal lines should be included only if they help to guide the reader's eye;

—Vertical lines are seldom necessary.

Rule 6: Give a verbal summary of the main points of the table
This is the most important rule. It is essential that a clear and succinct summary should accompany each table. The rule demands that the writer of the report should carefully consider the role of the table: what is the current argument? How does the table contribute to it? What should the reader remember after he or she has scanned the table?

The verbal summary should contain only a few key points (say three or four); it should never be a blow-by-blow account of each entry in the table. It is usually a mistake to include explanations about changes of definition in the verbal summary. In general these belong as footnotes although there may be occasions when it is desirable to emphasise that an apparently interesting change in a trend is due to a change in definition: if this is thought necessary the explanation should appear at the end of the verbal summary rather than as the first or second point.

Clearly the emphasis given to patterns and trends in any table will be dictated by the role of the table in a particular report and on the expected readership. This is inevitably a subjective decision and so it is important to consider the table carefully so as to ensure that your summary is an honest exposition of the data. For example, in Table 4.16, even if your audience is likely to be particularly interested in public transport vehicles, it would not be appropriate to dwell at length on the increase in that category of vehicle (from 94 to 112 thousand) without pointing out the much more dramatic change in the number of private cars and vans licensed.

On the assumption that Table 4.16 is intended for a general report on trends in the numbers of licensed vehicles, the verbal summary which accompanied the original table in *Social Trends* 13 is clear and well-balanced. It says:

'The number of licensed private cars and vans in the United Kingdom increased by 156 per cent between 1961 and 1981, from 6.1 to 15.6 million [table 4.12]. Most of this increase took place during the 1960s—the number doubled between 1961 and 1971. On the other hand, the number of motorcycles, scooters and mopeds fell between 1961 and 1971, but increased by about a third to 1.4 million in 1981; however, this was still under the 1961 level of 1.8 million.'

. . . a clear, concise summary of the two main patterns in the table. The table is specifically referred to by its number in order to link it in with the text; the numbers on which the summary statements are based are quoted (. . . the number 'increased by about a third to 1.4 million in 1981'); and no attempt is made to say something about each category in turn.

Minor alterations could be suggested: for example it would be possible to comment on the increasing dominance of private cars and vans (from about 60 per cent of all vehicles in 1961 to over three-quarters of all vehicles by 1971) or to comment on the fact that the total number of vehicles licensed almost doubled over the period from 10.2 to 19.8 million). But essentially the verbal summary is well designed to help an interested but non-specialist audience appreciate the main patterns in the table.

It is a useful discipline to write the verbal summary *before* deciding on the final design of table to include in a report. This will help to ensure that the figures most frequently compared with each other are shown in columns as well as ensuring that the statistics quoted in the table are those which most effectively illustrate the main patterns in the data.

Writing a verbal summary will also help to decide how many separate categories should be shown in the final table. Many tables include a final category labelled 'others' or 'miscellaneous'. In Table 4.16 the 'other vehicles' category is larger than the 'public transport vehicles' category and, in 1981, larger than the 'agricultural tractors' category. In such circumstances it is reasonable to consider absorbing these two relatively small categories into an enlarged 'others' category. Clearly this cannot be done if the trend in either of these categories is commented on in the verbal summary —either because of its intrinsic interest or because of the known interests of the expected readers of the report.

It will not always be immediately obvious where minor categories can be safely amalgamated: their pattern of growth or decline must be compared with trends in other columns of the table—best done by applying rules 1 to 4

on the whole table—in order to establish whether or not a comparatively small category merits specific mention in the verbal summary. However, when nothing particularly noteworthy has been observed in minor categories, the final table will be more compact and the main patterns clearer if these small categories are absorbed into an expanded 'others' category.

Since the proposed verbal summary makes no mention of either 'public transport vehicles' or 'agricultural tractors', and since when these two categories are included with the 'others' the expanded 'other vehicles' category is still considerably smaller than any of the three major categories, the amalgamation seems sensible, and the final table with an alternative verbal summary is given below.

Verbal summary to Table 4.17
Table 4.17 shows that the number of licensed vehicles almost doubled between 1961 and 1981, from 10.2 million to 19.8 million. This increase was almost entirely due to the growth in private cars and vans, from 6.1 million in 1961 to 15.6 million in 1981. By 1981, 79 per cent of licensed vehicles were private cars or vans compared with 60 per cent in 1961. Most of this growth took place in the 1960s. On the other hand, the number of motorcycles, scooters and mopeds fell sharply during the 1960s (from 1.8 million in 1961 to one million in 1971) but since then it has increased, reaching 1.4 million by 1981.

Rule 7: Use charts to show relationships
Charts are excellent for conveying broad trends and relationships: 'the rate of growth increased sharply in 1975'; or 'in 1970 categories A and B were roughly the same size: by 1982 category B completely dominated category A'; or 'the percentage of liquid steel made using

Table 4.17 Motor vehicles currently licensed: simplified version of table

United Kingdom 1961 to 1981   *Thousands*

Year	Type of Vehicle				All Vehicles
	Private cars and vans	Goods vehicles[1]	M'cycles, scooters, mopeds[2]	Public transport, agric. tractors, others[3]	
1961	6,100	1,490	1,800	780	10,200
1966	9,700	1,610	1,400	830	13,600
1971	12,400	1,660	1,000	810	15,900
1976	14,400	1,800	1,200	410	18,200
1981[4]	15,600	1,800	1,400	1,000[5]	19,800
Average	11,600	1,670	1,390	850	16,400

Source: Department of Transport

[1] Includes agricultural vans and lorries, tower wagons, showmen's goods vehicles and also vehicles licensed to draw trailers. Also includes Northern Ireland figures for general haulage tractors.
[2] Also includes Northern Ireland figures for three wheelers (under 406 kilos) and pedestrian controlled vehicles.
[3] Excludes Northern Ireland. Includes taxis, three wheelers, pedestrian controlled vehicles and general haulage contractors. Tram cars are included in 1961 and 1966.
[4] From 1978 the census has been taken on 31 December: previous censuses were taken on 30 June.
[5] Includes 170 thousand vehicles in exempt tax classes 61 and 62 which were not included in previous censuses.

the open hearth process decreased markedly during the 1960s'; or 'series A and B declined at similar rates during the 1970s: during the same period, series C grew rapidly, overtaking both series A and B by 1976'.

Messages like these can all be illustrated effectively using charts: a line graph for the first; grouped bar charts for the second: pie charts for the third and three line graphs for the fourth. But in all cases the reader will be left with a general impression rather than a memory of any specific amounts. This may be completely appropriate. If you find that your verbal summary includes a number of qualitative statements that do not depend on particular numerical values, you should pause and consider whether the main patterns in your data might be communicated more effectively using a chart rather than a table. Chapters 6 and 7 are devoted to the subject of statistical charts.

However, even where general statements are included in the verbal summary, the data may not lend themselves to a graphical presentation. For example, consider Table 4.17 and the verbal summary underneath it. The three points singled out for comment are:

1. the increase in the total number of licensed vehicles;

2. the increase in the number of private cars and vans;

3. the decline and subsequent recovery in the number of motorcycles etc.

It is extremely difficult to devise a single chart which illustrates these three points as effectively as the table. If the total number of vehicles is represented by a series of vertical bars (as in Figure 4.1), the overall pattern of growth can be seen: but any attempt to subdivide these bars into four slices of changing sizes as in Figure 4.2 leads to a muddled impression, and any change in the number of motorcycles from year to year is almost impossible to distinguish. If the four categories are plotted as line graphs against time as in Figure 4.3, we find that the scale needed to include the private vehicles numbers on a compact chart, reduces the motorcycle line to a low horizontal wiggle. This is hard to distinguish from the goods vehicles line or the public transport line.

Figure 4.1 Motor vehicles currently licensed United Kingdom 1961 to 1981

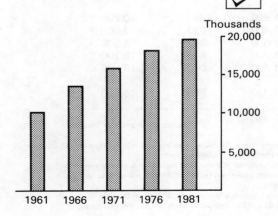

Figure 4.2 Motor vehicles currently licensed United Kingdom 1961 to 1981

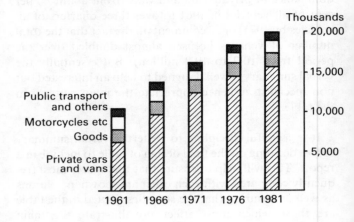

Figure 4.3 Motor vehicles currently licensed United Kingdom 1961 to 1971

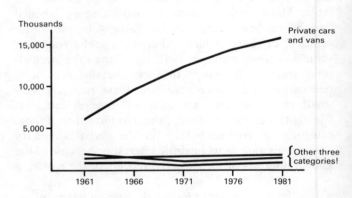

These arguments are not meant to deny the value of statistical graphs and diagrams: for some data displays good charts are much more effective than tables. However, some sets of data are unsuitable for graphical presentation because of the scales involved, and if the reader is invited to register specific quantities charts are likely to be less effective than well-designed tables.

Finally it is worth repeating that demonstration tables should appear in the body of a report. Very few readers will take the trouble to flip backwards and forwards through a report looking for tables in an annex—and so very few readers will pay any attention to the quantitative basis of your argument. (If the figures don't matter, leave them out.) This means that tables should be compact enough to be included in the text. They should also appear as close as possible to their verbal summary, either on the same page or on a facing page so that the table can be consulted easily while the verbal summary is being read. This proximity is much easier to achieve if you use several small demonstration tables, each illustrating only two or three points rather than a single large table.

Occasionally, when the message to be conveyed is not a simple one, it is right to venture beyond the rules and guidelines given in this chapter and to *dramatise* a table by turning it into a picture. An example, previously given in a

Royal Statistical Society Paper,[1] comes from some manpower planning work done some years ago for a group of Civil Service specialists; let us call them philosophers. The point to be got across was that the grade structure which would have held in 1980 if staffing estimates proved accurate was incompatible with a specified career prospectus which was thought to be desirable (though not necessarily attainable). It was taken as axiomatic that all posts in the top four grades should be filled by 'permanent' philosophers but that some 'transient' people (who spend just a few years in philosophy before moving on to other occupations) would be acceptable in the two bottom grades.

Table 4.18 was put in a draft report but some readers found it hard to understand. One reader eventually puzzled it out by drawing himself a diagram. What he drew is shown in Figure 4.4. His diagram was used in the final report in place of the original table. It shows with clarity what the original table shows only cryptically; it brings the figures to life and is superior in every way.

No rules can be given on how to dramatise tables; it is largely a matter of improvisational flair (which, in our example, was provided by a non-statistician). What can

Table 4.18 Estimated number of posts which would be needed in 1980 to maintain the desirable career prospectus for Philosophers

	Permanent*	Transient
Director	56 (20)	
Deputy Director	81 (11)	
Superintendent	439(177)	
Chief	721	
Senior	1,122	880
Basic	1,847	1,746
Total	4,226(208)	2,626

* Figures in brackets are number of posts in excess of the staffing estimates for 1980.

be said, however, is that such unorthodox methods should only be used in particular cases where they offer clear advantage over orthodox tables (or charts) and that they are likely to be much more effective when used in full knowledge of the rules and guidelines set out in this and other chapters.

Figure 4.4 Career implications of 1980 staffing estimates for Philosophers

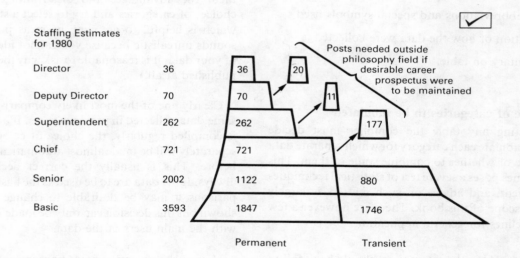

Staffing Estimates for 1980		Permanent	Transient
Director	36	36	20
Deputy Director	70	70	11
Superintendent	262	262	177
Chief	721	721	
Senior	2002	1122	880
Basic	3593	1847	1746

Posts needed outside philosophy field if desirable career prospectus were to be maintained

[1] MAHON, B. Statistics and decisions: the importance of communication and the power of graphical presentation. *Journal of the Royal Statistical Society, Series A*; general, vol 140, pp 298–323.

Chapter 5: Reference tables

5.1 Introduction

The criterion for a good reference table is extremely simple: it should be easy to use. This means that the reader should have no difficulty in identifying the row and/or column which is wanted, no difficulty in reading along a row or down a column to the appropriate entry, no difficulty in finding the exact definition of any entry and no doubt about where, when and how the data were collected.

Many of the guidelines which are important in designing good reference tables have already been discussed in Chapter 3. Points of style which need special consideration for reference tables are:

—which categories to show separately

—which factor to show in rows and which in columns

—ordering of rows and columns

—layout of large tables

—treatment of errors

—footnotes and general explanatory notes covering more than one table

—key to abbreviations and special symbols used

—explanation of how the data were collected

—commentary on table.

5.2 Choice of categories to be tabulated

In constructing any table the compiler must decide whether to tabulate each category for which separate data are available or whether to combine some of them. This task sometimes poses a severe test of statistical techniques and judgement, and advice on such matters is largely outside the scope of this book. There are however a few simple guidelines for general application.

In reference tables, the category set should be as full as is likely to be required by users, consistent with cost constraints and reasonable size of tables. Sub-totals should be provided where appropriate and, if the primary data allow it, the chosen categories should conform to a standard set, such as those in the Standard Industrial Classification (SIC).

The advantage of using a standard set of categories is that the data can then be compared directly with other data collected according to the same classification. This is an important point as data in reference tables will frequently be used for further analysis involving wider comparisons.

Unfortunately not all standard sets of categories are compatible with each other: for example some, but not all of the categories defined in the Standard Industrial Classification coincide with those used in the Standard International Trade Classification (SITC). (At the highest level of aggregation, the SIC division 'Other Manufacturing Industries' includes food, drink and tobacco manufacturing industries: in the SITC the section 'Manufactured Goods' omits food, drink and tobacco. At more detailed levels of aggregation there are many minor differences in definition between categories which at first sight, appear the same.)

Similarly if you categorise data on the social class of respondents according to the Registrar General's socio-economic grouping your data will not be directly comparable with tables in which respondents are classified according to the Institute of Practitioners in Advertising (IPA) categories. And vice versa.

In cases like this, the only sensible approach is to consider carefully the likely needs of users of your table(s). If possible ask them what analyses they carry out on the data and what other data sources they use. Discuss with them the advantages and disadvantages of different choices of categories and try to select a set of categories which is helpful to as many users as possible. (If this sounds unrealistic because you cannot identify the users of your data, it is reasonable to ask why the data are being published at all.)

Clearly one of the most likely comparisons is with the same data collected in previous years. If a reference table is compiled regularly, the choice of categories to show separately will be made almost automatically: the same as before. This is usually the correct decision, but not always. If the data are to be used as the basis for new comparisons it may be desirable to change the categories shown but this decision can only be made in consultation with the main users of the data.

Having chosen categories to be represented explicitly in the table, one is usually left with a remainder or 'other' category. Whenever possible the 'other' category should be smaller than the smallest category shown separately. In this flawed world there is often also an 'unknown' category. It is bad practice to combine these two into an 'other and unknown' category except when the combined category is small.

5.3 Which factors to put in rows and which in columns

It is easier to scan down a column in search of a particular item than to search for it along a row. This means that, in

Designing a reference table, you must again put yourself in the user's position to decide how the table should be oriented.

Consider, for example, Table 5.1 giving the index of industrial production from 1973 to 1981. A user of this table is unlikely to extract a single number from the table.

Table 5.1 Index of industrial production (Detailed analysis)

Average 1975=100

	Mining and quarrying			Food, drink and tobacco		Chemicals, coal and petroleum products		Metal manufacture		Engineering and allied industries				
										Engineering				Ship-building and marine engineering
	Coal mining	MLH 104[1]	Other mining and quarrying	Food	Drink and tobacco	Coal and petroleum products	Chemicals and allied industries	Ferrous	Non-ferrous	Total engineering	Mechanical engineering	Instrument engineering	Electrical engineering	
MLH[2]	101	104	102, 103 109	211-229	231-240	261-263	271-279	311-313	321-323	331-369	331-349	351-354	361-369	370
Weights	34	0·3	7	54	23	9	57	35	12	170	92	12	66	14
1973	103·2		121·1	106·0	96·3	120·6	105·3	129·5	117·0	98·4	97·1	92·4	101·2	95·4
1974	85·4		112·2	103·9	99·1	116·0	109·5	116·2	113·3	102·3	101·2	98·7	104·5	98·9
1975	100·0	100	100·0	100·0	100·0	100·0	100·0	100·0	100·0	100·0	100·0	100·0	100·0	100·0
1976	92·7	5 123	96·3	103·0	101·3	105·7	112·0	104·7	105·4	96·6	95·4	97·7	98·1	96·5
1977	90·1	15 023	95·8	105·2	101·0	102·7	115·9	103·1	108·3	97·5	94·0	100·1	102·0	93·5
1978	89·2	21 796	103·9	106·3	106·1	101·4	117·0	102·3	106·9	99·4	92·4	106·2	107·9	86·4
1979	89·3	31 333	106·0	107·3	108·4	105·3	119·3	104·4	105·3	101·2	91·3	109·4	113·6	78·1
1980	91·9	32 110	97·4	106·5	106·9	93·3	109·9	67·1	97·3	97·6	85·4	107·1	112·9	67·8
Seasonally adjusted														
1976														
1st quarter	94·8	2 400	98·4	100·0	95·5	102·3	107·7	100·3	101·2	96·1	96·7	95·3	95·3	96·9
2nd quarter	93·8	4 100	98·9	103·7	102·5	105·1	110·7	109·7	105·1	97·3	96·7	96·8	98·2	93·4
3rd quarter	91·1	5 300	94·1	104·5	105·6	105·3	113·4	106·1	106·1	96·0	94·2	96·5	98·4	92·7
4th quarter	91·3	8 600	93·8	103·8	101·7	110·1	116·4	102·6	109·2	97·2	94·1	102·2	100·6	103·0
1977														
1st quarter	91·2	12 800	100·0	106·8	101·1	105·6	117·9	106·8	113·1	98·8	96·9	98·7	101·4	92·1
2nd quarter	91·0	15 400	94·5	104·2	98·5	103·4	115·8	103·5	107·7	96·2	93·0	100·2	99·9	96·5
3rd quarter	89·7	15 500	94·6	105·1	100·4	101·6	116·0	105·1	105·7	97·6	93·6	100·9	102·6	95·9
4th quarter	88·4	16 500	94·1	104·7	104·0	100·2	113·9	97·0	106·9	97·6	92·6	100·6	104·1	89·2
1978														
1st quarter	89·5	18 500	95·1	104·4	105·3	98·9	114·4	107·1	105·1	98·2	92·4	102·3	105·5	89·5
2nd quarter	89·1	21 400	102·0	108·2	107·8	99·2	116·7	107·0	108·6	98·9	92·1	105·7	107·2	87·9
3rd quarter	89·4	22 200	105·8	107·7	103·3	104·6	118·8	97·0	107·1	100·6	94·0	106·5	108·8	85·7
4th quarter	88·7	25 000	112·8	104·6	108·1	102·9	118·3	98·0	106·7	99·8	91·1	110·2	110·2	82·6
1979														
1st quarter	88·0	29 300	96·4	105·9	105·2	101·2	114·9	98·5	105·0	99·8	89·8	107·1	112·4	82·4
2nd quarter	87·5	31 800	110·3	107·2	109·3	103·4	122·3	112·8	107·6	104·2	94·4	110·7	116·8	81·9
3rd quarter	90·8	33 000	110·3	108·3	109·7	109·3	121·0	103·2	103·8	97·6	87·6	108·2	109·7	75·9
4th quarter	91·0	31 200	107·0	107·8	109·4	107·2	118·7	103·0	104·8	103·3	93·6	111·5	115·5	72·1
1980														
1st quarter	92·4	32 600	107·8	106·9	114·7	98·8	119·2	41·4	104·7	103·7	93·0	110·4	117·6	70·3
2nd quarter	90·7	31 700	99·7	105·5	104·4	93·4	110·1	91·5	102·5	98·2	86·8	107·8	112·3	68·1
3rd quarter	91·4	30 700	93·9	105·8	103·7	90·3	105·4	70·3	95·1	97·1	83·7	109·0	113·6	66·7
4th quarter	93·0	33 500	88·2	107·9	104·9	90·7	104·7	65·1	87·0	91·5	78·2	101·4	108·2	66·1
1981														
1st quarter	90·4	35 000	88·2	106·2	108·7	86·0	105·7	73·1	86·2	89·7	78·4	98·7	103·8	65·1
2nd quarter	90·7	34 500	90·1	104·5	96·1	83·4	108·7	77·4	84·7	89·2	76·3	97·3	105·8	66·0
1980														
August	90		94	105	101	90	102	69	100	97	84	109	114	66
September	92		94	107	107	86	106	66	88	96	82	108	113	67
October	93		88	109	107	89	106	63	89	93	79	103	110	67
November	93		88	106	105	91	102	65	94	91	78	101	108	66
December	93		89	108	103	91	106	68	78	90	77	99	106	66
1981														
January	92		88	107	110	89	104	69	87	90	78	99	104	66
February	89		88	106	117	85	107	74	87	90	78	98	103	65
March	90		88	106	99	84	106	76	84	90	78	98	104	65
April	90		89	105	98	89	107	72	86	89	77	97	105	66
May	91		92	105	97	82	105	75	88	89	76	97	106	65
June	91		89	104	93	78	115	85	81	89	76	98	107	67
July	90		90	103	100	81	112	74	85	90	76	100	107	67
August	89		89	107	104	82	114	75	89	90	..	..	..	68

Source: Central Statistical Office

1. MLH 104: extraction of mineral oil and natural gas.
2. Industries are grouped according to the Standard Industrial Classification 1968.
.. Not available.

49

He or she is most likely to want to study the trend over time within one category. Thus, having found the column (or row, if the table had been produced the other way round) of interest, the user will typically copy down several numbers from this column. This extraction is easier if time is shown vertically, as in Table 5.1.

In Table 5.2 (from *Employment Gazette*, November 1982) which records unemployment by age and duration, users are likely to compare the distribution of durations of unemployment for different age groups. This entails first identifying the age group to be studied (column headings), then for each age group in turn recording the number of people who had been unemployed for under one week, over one and up to two weeks, and so on: in other words finding and copying down a number of items from the same column. Thus, for each search along a row which is comparatively slow, the user will carry out a number of vertical searches which are quicker and easier.

The principle is that the numbers bearing the most clearly identifiable relationship to one another should be put in columns.

A related principle is that it is generally better to have figures of about the same size arranged in columns rather than rows, because this results in a neater and easier to use table.

Occasionally these two principles conflict but more often than not they reinforce one another. For instance, when time is one of the factors in a table, both principles call for the time categories to be arranged vertically. Demonstration of the improvement brought about by

such an arrangement is provided by Tables 3.8, 3.9 and 3.10 (pages 25 and 26).

5.4 Order of tabulation of categories

Some categories, such as time periods, have their own natural order but others leave you free to choose in which order to put them in the table.

In tables, use the relevant standard order, for example, alphabetical order or that of the Standard Industrial Classification. If no standard order exists, decide your own order and stick to it in future tables so that related figures from tables for different years can be easily compared or extracted.

It is generally good practice to put the 'unknown' category last and the 'other' category second last if these two categories appear in a table.

As usual, the important principle is to think carefully about the order of categories and make a deliberate choice with the interests of users at heart.

5.5 Layout of large tables

In general, reference tables should be printed vertically on the page rather than horizontally. However, if a number of reference tables are all wide and shallow, it will be convenient for the whole set to be printed horizontally. See, for example, Table 5.3 which is one of a large set of tables giving statistics on further education in 1980.

An alternative layout for very wide tables is to extend them across two pages. When this is done the title on the

Table 5.2 Unemployment by age and duration: October 14 1982

Duration of unemployment in weeks United Kingdom		Under 18	18	19	20-24	25-29	30-34	35-44	45-49	50-54	55-59	60-64	65 and over	All
MALE														
One or less		4,555	2,357	2,019	8,162	5,182	3,985	5,752	2,228	2,015	2,054	1,891	19	40,219
Over 1 and up to	2	6,449	3,801	2,976	12,714	7,989	6,113	8,502	3,505	3,368	3,932	3,960	41	63,350
2	4	12,220	7,666	5,647	21,990	12,845	9,871	14,129	5,504	5,109	5,282	5,187	43	105,493
4	6	30,863	12,938	5,991	26,049	12,791	9,421	12,917	5,156	5,168	5,645	5,039	35	132,013
6	8	8,999	6,542	4,411	23,402	11,046	7,987	11,097	4,450	4,433	5,039	5,013	37	92,456
8	13	18,030	12,671	9,522	34,340	21,657	17,225	24,410	9,908	9,386	11,025	11,216	82	179,472
13	26	42,925	24,237	18,327	63,421	39,859	31,666	44,215	18,110	18,541	22,978	25,912	162	350,353
26	39	12,124	15,094	12,496	42,903	29,429	24,062	34,107	13,874	14,795	19,078	22,939	182	241,083
39	52	6,072	9,174	10,198	34,028	23,578	19,319	27,827	11,651	12,586	17,141	24,078	236	195,888
52	65	6,058	7,112	9,722	30,775	20,351	16,510	23,025	9,615	10,634	14,999	21,507	257	170,565
65	78	3,000	4,723	7,043	23,276	16,350	13,708	19,445	8,335	9,333	13,294	19,623	268	138,398
78	104	782	4,330	9,624	37,558	28,341	24,260	35,015	14,748	16,124	20,661	29,786	653	221,882
104	156	198	2,785	6,435	42,739	30,387	25,840	36,408	15,895	17,003	19,436	27,351	746	225,223
156		—	135	906	15,325	16,156	16,223	29,340	15,996	19,296	22,622	24,853	1,413	162,265
All		152,275	113,565	105,317	416,682	275,961	226,190	326,189	138,975	147,791	183,186	228,355	4,174	2,318,660
FEMALE														
One or less		3,605	2,048	1,536	5,056	2,579	1,503	1,970	833	705	532		34	20,401
Over 1 and up to	2	5,462	3,432	2,369	8,305	4,160	2,510	3,203	1,351	1,201	1,005		38	33,036
2	4	10,425	6,992	4,464	14,018	6,924	4,202	5,555	2,243	1,882	1,415		70	58,190
4	6	26,347	14,156	4,863	18,039	7,557	4,629	5,929	2,213	2,060	1,707		66	87,566
6	8	7,306	5,736	3,550	15,702	5,865	3,452	4,575	1,898	1,773	1,636		49	51,542
8	13	14,067	10,271	7,002	21,063	12,214	7,439	9,613	3,873	3,432	2,903		117	91,994
13	26	34,344	19,893	13,808	40,907	23,360	14,158	18,306	7,820	7,385	6,380		238	186,599
26	39	8,393	9,948	8,521	26,939	17,877	10,663	13,741	6,306	6,036	5,414		212	114,050
39	52	4,489	5,923	6,627	20,191	13,147	7,798	9,405	4,568	4,740	4,767		210	81,865
52	65	4,413	4,823	6,253	16,043	9,101	5,594	7,528	3,719	4,147	4,658		199	66,478
65	78	2,165	2,995	4,088	9,933	4,905	3,310	4,844	2,747	3,240	3,472		173	41,872
78	104	563	2,565	4,735	12,906	6,713	4,526	7,039	4,047	5,041	5,650		310	54,095
104	156	159	1,648	3,345	15,003	6,348	4,104	6,620	4,139	5,455	6,081		398	53,300
156		—	104	705	7,009	4,088	2,656	4,408	3,263	5,124	7,544		579	35,480
All		121,738	90,534	71,866	231,114	124,838	76,544	102,736	49,020	52,221	53,164		2,693	976,468

Source: Employment Gazette, November 1982

Table 5.3 Age[1] of students at major establishments of further education by qualification aim, 1980

England and Wales

Course Enrolments[2]

					Men and Women				
	16	17	18	16-18	19	20	21-24	25 & over	Total
Advanced courses									
Teacher training									
Postgraduate initial		1		1	1	51	3,662	1,903	5,623
Other initial		69	2,623	2,692	4,551	5,268	6,066	5,996	24,573
Postgraduate in service						2	77	1,467	1,546
Other in service	1	1	3	5	2	6	687	15,573	16,273
Other advanced									
Higher deg, P-G, research					27	135	4,207	11,415	15,784
University first degree		47	1,969	2,016	2,676	2,470	2,008	2,013	11,183
CNAA first degree	5	280	12,745	13,030	18,405	18,256	31,548	18,001	99,240
HND, HNC	6	107	2,614	2,727	5,005	5,518	10,081	5,234	28,565
TEC/BEC higher cert/dip	111	288	7,498	7,897	13,247	9,140	13,702	8,424	52,410
Dip HE		4	227	231	351	218	547	1,321	2,668
Professional quals	92	248	3,760	4,100	6,443	7,076	28,019	34,305	79,943
College dip/cert	33	92	1,489	1,614	1,571	1,303	3,935	5,374	13,797
Other advanced	12	19	572	603	452	582	2,406	3,987	8,030
Total advanced	260	1,156	33,500	34,916	52,731	50,025	106,945	115,013	359,635
Non-advanced courses									
OND, ONC	2,395	3,566	2,344	8,305	1,718	1,205	1,609	458	13,295
TEC/BEC cert/diploma	37,616	43,463	32,831	113,910	20,379	11,581	16,690	15,719	178,279
City and Guilds	73,908	84,801	67,156	225,865	42,165	16,397	17,649	38,437	340,513
GCE A level	17,262	23,462	21,914	62,638	11,359	7,393	15,191	24,935	121,516
GCE O level	37,888	25,544	11,484	74,916	7,393	5,682	15,425	42,486	145,902
CSE, CEE	754	91	17	862	5	10	7	65	950
Professional quals	23,837	22,325	13,480	59,642	8,599	5,924	16,003	45,669	135,837
College dip/cert	2,556	3,118	5,011	10,685	1,899	1,144	2,246	3,464	19,438
Other specified	531	1,252	2,119	3,902	1,383	897	1,240	1,442	8,864
Unspecified	54,137	33,816	16,889	104,842	13,332	11,239	40,005	382,633	552,051
Total non-advanced	250,884	241,438	173,245	665,567	108,232	61,472	126,065	555,309	1,516,645
Total adv & non-advanced	251,144	242,594	206,745	700,483	160,963	111,497	233,010	670,327	1,876,280

[1] Age at 31 Aug
[2] All modes of attendance

initial page should be repeated on the continuation page followed by '(*continued*)' in italic type.

A double page spread may be the best way to present a table spanning two pages but care needs to be taken with this form of table because it requires precise alignment of the two pages by the printer—not always easy to achieve.

Where a table spans two pages it may be useful to repeat the row headings at the righthand side of the table. This is particularly desirable when there are large numbers of rows close together and it may be difficult to read across all the columns accurately.

5.6 Errors: indication of size of intrinsic error

Figures in tables are rarely accurate to the last digit. Almost every figure in almost every table has associated with it a range of probable error. Errors may arise at each stage of the process of data collection and storage: the original data may be incomplete, either by accident (non-return) or design (if the data were collected from a sample survey); the data may have been inaccurately recorded or transcribed; some data may have been assigned to incorrect categories; and simple arithmetic mistakes may have occurred in calculating totals or percentages. These factors all contribute to the intrinsic lack of accuracy of published data.

Table users should ideally be given some idea of the size of intrinsic errors in the figures, and yet the great majority of tables contain no such indication.

Why is this? There are two main problems. First, the size of the error may not be known, even to the person who collected the data. This is so particularly when the data collection system relies heavily on the truthfulness and reliability of people. Secondly, even when the probable size of error is known (for example, in the results from a properly designed and well-regulated sample survey) there is no convenient and simple means of showing it. Broadly speaking, you can do one of four things when compiling tables:

1. Ignore the existence of intrinsic errors altogether. This is often done by describing the figures in the tables as numbers 'recorded' without any reference to how closely these correspond to the numbers of *actual* objects, events, etc which they represent;

2. Make a general disclaimer covering the whole set of tables, saying, for example, that figures are not necessarily accurate to the last digit shown;

3. Indicate the size of errors implicitly by rounding figures to an appropriate degree;

4. Indicate the size of errors explicitly, by showing confidence intervals[1] or standard errors[1] or by giving a 'quality label' to each figure or set of figures (eg label A could indicate ranges of error of less than ± 1 per cent, B those between ± 1 per cent and ± 10 per cent and C those greater than 10 per cent).

The choice of 1., 2., 3., or 4. depends on circumstances. One sensible strategy is to adopt course 3. unless there are particular reasons for doing otherwise. It is important to realise however that when two or more rounded figures are used in a subsequent calculation the rounding errors (which may be insignificant in single numbers) can mount up to such an extent that the result of the calculation may be significantly wrong. The general rule is therefore that figures should be rounded sufficiently to cut out spurious accuracy but no more. It must also be borne in mind that rounding involves an extra process (which must be done *after* checking that the *original* figures sum to the *original* row and column totals) and therefore extra cost. This extra cost may not be justified in all circumstances.

Some possible reasons for dealing differently with the problem of intrinsic errors are as follows:

Reasons	Course to be adopted
Insufficient information exists about size of error. (To round the figures one needs to know at least that they are unlikely to be accurate to within so many per cent)	Course (2) or, occasionally, if all readers are known to be aware that errors exist, course (1).
Rounding may be uneconomical or impracticable for some reason	
Errors are known to be negligible	Course (1).
Probable error sizes are known and are sufficiently important to the users of the table to be stated explicitly	Course (4).

The main disadvantage of course (4) is that it can lead to a very complicated presentation, as for example, in Table 5.4 from the Department of Employment's *Family Expenditure Survey*, 1978.

5.7 Footnotes and explanatory notes covering a set of tables

The prime objective of footnotes and explanatory notes is to prevent misuse or misinterpretation of the data. They should therefore be used to record:

—any change in the coverage of entries in the table (eg 'Figures for 1981 include tax exempt classes 61 and 62, previously omitted from the censuses.');

—any change in the definition of terms used in the table (eg 'In October 1980 all direct grant secondary schools in England and Wales were reclassified as independent schools.');

---

[1] A full definition of these terms can be found in most statistical text-books but all the non-technical reader of this book needs to know is that a 95 per cent confidence interval is a range which is 95 per cent certain to include the true value and that the extreme points of such a range are usually about two standard errors above and below the central value.

Table 5.4 Expenditure of households at different levels of household income
(Family Expenditure Survey)

	Gross normal weekly income of household									
	£15 and under £20	£20 and under £25	£25 and under £30	£30 and under £40	£40 and under £50	£50 and under £60	£60 and under £80	£80 and under £100	£100 or more	All* house-holds
Total number of households	88	238	324	197	122	100	157	102	109	1,444
Average number of persons per household										
Males	0.227	0.231	0.188	0.264	0.336	0.360	0.548	0.422	0.578	0.319
Females	0.773	0.769	0.812	0.736	0.664	0.640	0.452	0.578	0.422	0.681
Adults										
Persons under 65	0.375	0.218	0.151	0.315	0.541	0.650	0.790	0.853	0.899	0.445
Persons 65 and over	0.625	0.782	0.849	0.685	0.459	0.350	0.210	0.147	0.101	0.555
Persons working	0.125	0.067	0.077	0.269	0.500	0.680	0.834	0.873	0.862	0.382
Persons not working	0.875	0.933	0.923	0.731	0.500	0.320	0.166	0.127	0.138	0.618
Men 65 and over, women 60 and over	0.705	0.824	0.873	0.685	0.467	0.260	0.159	0.127	0.110	0.561
Others	0.170	0.109	0.049	0.046	0.033	0.060	0.006	—	0.028	0.057
Average age of head of household	65	70	72	67	59	54	50	46	46	62
Housing by type of tenure	*Number of households*									
Rented unfurnished	76	169	235	85	40	37	67	29	15	756
Local authority	51	118	194	62	27	26	43	20	6	547
Other	25	51	41	23	13	11	24	9	9	209
Rented furnished	6	7	7	13	20	19	27	16	7	125
Rent-free	1	5	9	12	4	5	7	4	3	50
Owner-occupied	5	57	73	87	58	39	56	53	84	513
In process of purchase	—	5	2	5	6	9	13	28	54	122
Owned outright	5	52	71	82	52	30	43	25	30	391

Commodity or service	Average weekly household expenditure (£)									
Group totals										
Housing	3.52	4.47	7.63	9.22	9.33	10.59	10.04	11.56	15.68	8.57
Percentage standard error	11.2	4.1	4.1	6.8	6.0	10.3	5.3	6.8	5.3	2.3
Fuel, light and power	2.61	2.68	3.21	3.20	3.17	3.18	2.95	3.46	4.18	3.13
Percentage standard error	7.3	4.5	5.1	5.5	7.4	11.3	6.3	6.5	9.7	2.3
Food	6.71	6.87	7.19	8.06	9.61	9.79	9.20	9.68	12.06	8.37
Percentage standard error	3.7	2.5	2.0	2.9	4.1	5.9	3.8	4.8	6.3	1.4
Alcoholic drink	0.47	0.50	0.44	1.01	1.63	1.77	2.84	2.65	4.22	1.43
Percentage standard error	24.4	18.4	15.8	15.7	15.2	17.4	12.9	13.8	13.3	5.8
Tobacco	0.72	0.70	0.54	0.66	1.36	1.63	1.75	1.32	1.61	1.01
Percentage standard error	17.0	11.7	12.5	14.1	13.4	14.5	10.6	17.9	15.6	4.8
Clothing and footwear	1.19	1.14	1.35	2.04	2.60	3.65	3.71	5.25	4.36	2.43
Percentage standard error	38.3	15.2	12.0	17.4	18.0	17.7	15.4	18.4	18.4	6.1
Durable household goods	1.14	0.80	1.24	1.76	1.89	3.60	2.26	3.64	10.45	2.42
Percentage standard error	39.4	19.7	38.9	46.5	44.3	28.7	22.7	30.9	32.4	13.9
Other goods	1.37	1.59	1.69	2.30	2.61	3.91	2.95	3.50	5.03	2.49
Percentage standard error	8.0	5.5	5.4	7.0	7.7	24.1	8.3	9.1	11.9	4.0
Transport and vehicles	1.40	0.48	0.86	2.22	2.59	4.10	7.47	9.37	11.21	3.49
Percentage standard error	59.8	17.4	19.9	20.5	18.2	14.6	14.0	13.4	12.9	6.6
Services	2.13	2.00	2.04	3.35	4.21	5.34	5.36	9.92	10.70	4.21
Percentage standard error	21.0	10.3	6.1	9.2	11.6	15.4	9.6	13.6	13.8	4.7
Miscellaneous	[0.01]	[0.02]	[0.01]	0.18	[0.45]	[0.14]	0.45	0.53	[1.04]	0.25
Percentage standard error	66.7	55.0	50.0	33.9	78.0	43.8	37.8	27.6	59.6	24.5
Total, all expenditure groups	**21.27**	**21.24**	**26.18**	**34.01**	**39.43**	**47.70**	**48.97**	**60.86**	**80.53**	**37.80**
Percentage standard error	7.5	2.7	2.7	5.0	4.7	5.9	3.5	4.7	6.7	2.1

* Includes seven households with income below £15 not shown separately in this table.
[ ] This figure is based on 10 readings or less.

—further explanation of terms used in the table (eg 'Includes O level grades D and E, and CSE grades 2–5.');

—difference in status of some entries in the table (eg 'Figures from 1981 onwards are provisional.');

—explanation of conventions used in arriving at entries (eg 'The seasonally adjusted figures do not always add to the calendar year total which is the sum of unadjusted quarterly figures.').

Footnotes to reference tables must obviously be clear and complete. If you are responsible for the design of a set of related tables, the same footnote(s) may apply to a number of them. It is then tempting to introduce the set of tables with the explanatory notes which are common to all or several tables. This practice should be used conservatively. Many readers will refer to a single table in the middle of the set and may not bother to consult the beginning or end of the set of tables for covering notes. If they are in a hurry they may guess at the correct interpretation of an unclear entry rather than search for the appropriate explanatory note. It is therefore preferable to repeat the same footnote(s) after a number of tables rather than to preface a set of tables with general explanatory notes. If such notes are too extensive to be printed underneath a number of tables, then a concise footnote can be included referring users to the appropriate explanatory notes (eg 'See additional Notes 1, 2 and 5 on page XX.').

Footnotes may be used simply to amplify a row or column heading which was too cumbersome to print in full in the main table. For example, a column headed 'MLH 104' might be used with the accompanying footnote 'MLH 104: extraction of mineral oil and natural gas.' The physical arrangement of footnotes is discussed in Chapter 3, page 31.

5.8 Definitions

A list of definitions should be used to give precise explanations of terms which are used in a set of tables: for example, terms like 'Temporarily stopped workers', 'Weekly hours worked', and 'Short-time working' in tables of employment statistics. This list should also include definitions of any technical terms used in the tables and should normally appear with the entries in alphabetic order immediately after the tables. Figure 5.1 shows the list of definitions printed at the end of the tables in the Department of Employment's *Employment Gazette*.

5.9 Treatment of years

It is important to record the exact time period over which data have been collected. The 12-month period beginning 1 January 1984 is clearly labelled as 1984: however data are often recorded for financial years, academic years or other 12-month periods. In such cases, a consistent treatment should be used and should be explained to the

reader. A convention, commonly used in the Government Statistical Service, and adopted in this handbook is as follows:

1983–84 (with hyphen) denotes the financial year April 1983 to March 1984.

1983/84 (with oblique stroke) denotes the academic year September 1983 to August 1984.

1983–1984 (with hyphen and all four digits given for the second year) denotes the 2-year span January 1983 to December 1984.

5.10 Key to abbreviations and special symbols used

When appropriate, a key to abbreviations and symbols should be included either at the beginning or the end of any set of reference tables, preferably on a page by itself for immediate visibility. A typical list might look like this:

..	not available
–	nil or negligible (less than half the final digit shown)
e	estimated
n.e.s.	not elsewhere specified
p	provisional
MLH	Minimum List Heading of the SIC 1968
R	revised
SIC	UK Standard Industrial Classification 1968
EC	European Community

Obviously it is important to use the same symbols and abbreviations in all tables and, if possible, to choose ones which are compatible with those used in similar tables elsewhere.

5.11 Notes on method of data collection

Reference tables will be used by researchers who may need to know exactly how the data were collected: whether by a complete census or a sample survey; how the sample was selected; how respondents were stratified; what the response rates were in different subsamples; whether questionnaires were filled in by an interviewer or by the respondent; exactly how questions were worded and what, if any, prompts were used. If the data come from administrative records the researcher may need to know how the records are compiled and exactly how they have been used to produce the tabulated data.

If the table is a one-off production, it is probably best to include such information in a technical appendix. If the table is one of a regularly published series, all the necessary information about the methods used to collect the data should be easily accessible to researchers. This may be in the form of a separate booklet such as, for

Figure 5.1 Definitions taken from _Employment Gazette_ November 1982

DEFINITIONS

The terms used in the tables are defined more fully in periodic articles in Employment Gazette _relating to particular statistical series. The following are short general definitions._

BASIC WEEKLY WAGE RATES
Minimum entitlements of manual workers under national collective agreements and statutory wages orders. Minimum entitlements in this context means basic wage rates, standard rates, minimum guarantees or minimum earnings levels, as appropriate, together with any general supplement payable under the agreement or order.

DISABLED PEOPLE
Those eligible to register under the Disabled Persons (Employment) Acts 1944, and 1958; this is those who, because of injury, disease or congenital deformity, are substantially handicapped in obtaining or keeping employment of a kind which would otherwise be suited to their age, experience and qualifications. Registration is voluntary. The figures therefore relate to those who are registered and not those who, though eligible to register, choose not to do so.

EARNINGS
Total gross remuneration which employees receive from their employers in the form of money. Income in kind and employers' contributions to national insurance and pension funds are excluded.

EMPLOYED LABOUR FORCE
Total in civil employment plus HM forces.

EMPLOYEES IN EMPLOYMENT
Civilians in the paid employment of employers (excluding home workers and private domestic servants).

FULL-TIME WORKERS
People normally working for more than 30 hours a week except where otherwise stated.

GENERAL INDEX OF RETAIL PRICES
The general index covers almost all goods and services purchased by most households, excluding only those for which the income of the head of household is in the top 3–4 per cent and those one and two person pensioner households of limited means covered by separate indices. For these pensioners, national retirement and similar pensions account for at least three-quarters of income.

HM FORCES
All UK service personnel of ΗΜ Regular forces, wherever serving, including those on release leave.

INDEX OF PRODUCTION INDUSTRIES
SIC Orders II–XIX. Manufacturing industries plus mining and quarrying, construction, gas, electricity and water.

INDUSTRIAL DISPUTES
Statistics of stoppages of work due to industrial disputes in the United Kingdom relate only to disputes connected with terms and conditions of employment. Stoppages involving fewer than 10 workers or lasting less than one day are excluded, except where the aggregate of working days lost exceeded 100.

Workers involved and working days lost relate to persons both directly and indirectly involved (thrown out of work although not parties to the disputes) at the establishments where the disputes occurred. People laid off and working days lost elsewhere, owing for example to resulting shortages of supplies, are not included. There are difficulties in ensuring complete recording of stoppages, in particular those near the margins of the definitions; for example, short disputes lasting only a day or so. Any under-recording would particularly bear on those industries most affected by such stoppages; and would have much more effect on the total of stoppages than of working days lost.

MANUAL WORKERS
Employees other than those in administrative, professional, technical and clerical occupations.

MANUFACTURING INDUSTRIES
SIC Orders III–XIX.

NORMAL WEEKLY HOURS
The time which the employee is expected to work in a normal week, excluding all overtime and main meal breaks. This may be specified in national collective agreements and statutory wages orders for manual workers.

OVERTIME
Work outside normal hours for which a premium rate is paid.

PART-TIME WORKERS
People normally working for not more than 30 hours a week except where otherwise stated.

PENSIONER HOUSEHOLDS
Retail prices indices are compiled for one- and two-person pensioner households, defined as those in which at least three-quarters of total income is derived from national insurance retirement and similar pensions.

SEASONALLY ADJUSTED
Adjusted for regular seasonal variations.

SELF-EMPLOYED PEOPLE
Those working on their own account whether or not they have any employees.

SERVICE INDUSTRIES
SIC Orders XXII–XXVII.

SHORT-TIME WORKING
Arrangements made by an employer for working less than regular hours. Therefore, time lost through sickness, holidays, absenteeism and the direct effects of industrial disputes is not counted as short-time.

TEMPORARILY STOPPED
People who at the date of the unemployment count are suspended by their employers on the understanding that they will shortly resume work and are registered to claim benefit. These people are not included in the unemployment figures.

UNEMPLOYED
People registered for employment at a local employment office or careers service office on the day of the monthly count who on that day have no job and are capable of and available for work. (Certain severely disabled people, and adult students registered for vacation employment, are excluded).

UNEMPLOYED PERCENTAGE RATE
The number of registered unemployed expressed as a percentage of the latest available mid-year estimate of all employees in employment, plus the unemployed at the same date.

UNEMPLOYED SCHOOL LEAVERS
Unemployed people under 18 years of age who have not entered employment since terminating full-time education.

VACANCY
A job notified by an employer to a local employment office or careers service office.

WEEKLY HOURS WORKED
Actual hours worked during the reference week and hours not worked but paid for under guarantee agreements.

WORKING POPULATION
Employed labour force plus the registered unemployed.

Conventions The following standard symbols are used:

..	not available	e	estimated
—	nil or negligible (less than half the final digit shown)	MLH	Minimum List Heading of the SIC 1968
[]	provisional	n.e.s.	not elsewhere specified
——	break in series	SIC	UK Standard Industrial Classification 1968
R	revised	EC	European Community

Where figures have been rounded to the final digit, there may be an apparent slight discrepancy between the sum of the constituent items and the total as shown.
Although figures may be in unrounded form to facilitate the calculation of percentage changes, rates of change, etc. by users, this does not imply that the figures can be estimated to this degree of precision, and it must be recognised that they may be the subject of sampling and other errors.

example, the technical handbook published by HMSO[1] which describes the methods used to collect data in the Family Expenditure Survey. Alternatively a compact description may be included in the covering notes along with an address and telephone number for further enquiries, or an address and telephone number for enquiries can be given alone so long as up-to-date copies of a suitable document are readily available.

5.12   Commentary on reference tables

Strictly speaking, reference tables do not need a commentary. They provide data, well laid out and clearly explained, for users to extract and analyse according to their individual needs.

Nevertheless there are circumstances in which it is common to publish a commentary along with reference tables. These are:

a. in regular statistical publications, such as *British Business*[2] and *Employment Gazette*[3] where tables or sets of tables are usually introduced by a brief commentary on trends and changes revealed in the latest tables; and

b. when a specific investigation has been undertaken in order to provide reference data in a previously unquantified area, it is customary to accompany the tables with a summary of the main patterns revealed by the investigation. (For example, a survey conducted by the Office of Population Censuses and Surveys (OPCS) in order to find out how many mothers breastfed their babies and what factors affected the choice of breast or bottle feeding, might be used to provide data for future analyses and comparisons but the main findings of the survey would certainly be reported in some detail.)

The two circumstances are rather different. In the first it is reasonable to assume that the commentary is intended for regular users of this particular data and so all that is needed is a brief statement of the latest changes. In the second, the data have been collected in order to chart a previously unquantified area and the report should provide a clear overview of what was discovered as well as more detailed commentary on individual tables.

The first sort of example actually occurs when a single table is used both for demonstration and reference purposes. Although this approach is not recommended when writing for general readers, it may be quite acceptable in specialist publications where most readers are sufficiently familiar with the data to interpret the figures from a reference-type table.

In the second example it is quite possible for the overview report to include additional demonstration tables and charts, based on the reference tables but designed so as to highlight points of particular interest.

The subject of report writing is beyond the scope of this book but some guidelines on writing about numbers are discussed in Chapter 8.

5.13   Some further examples

Finally, it is instructive to examine some tables with a critical eye.

The first example has already appeared as Table 5.1, but it is repeated opposite as Table 5.5 for ease of reference. This table is an extract from the Index of Industrial Production table in the Central Statistical Office's *Monthly Digest of Statistics*. It is primarily a reference table but has been laid out so well, that, without compromising its main purpose, it also serves to demonstrate very effectively the changes in volume of production in each industry. The table heading and header detail can be criticised for skimping on such important items as the geographical coverage (is it GB or UK?) but the main body of the table in superb. It illustrates in particular the following points of style:

1. Horizontal ruled lines are kept to a functional minimum and there are no vertical lines.

2. Columns are evenly spaced irrespective of length of column headings. This has been achieved by spreading long words such as engineering over two lines and using a footnote to give one particularly long column title in full (MLH 104).

3. Character size and row and column spacing have been carefully chosen to enable as much information as possible to be clearly set out on each page.

4. Weights, being different entities from the index values shown in the rest of the table, are shown in italics.

5. In numbers greater than 1,000, a narrow space has been used, rather than a comma, to separate the thousands from the hundreds.

[1] KEMSLEY, W F F *Family Expenditure Survey: handbook on the sample, fieldwork and coding procedures.* Government Social Survey (M.146) HMSO, 1969

[2] *British Business:* a publication giving 'Weekly news from the Departments of Industry and Trade', published by HMSO

[3] *Employment Gazette:* a monthly publication from the Department of Employment, published by HMSO.

Table 5.5 Index of industrial production (Detailed analysis)

Average 1975=100

	Mining and quarrying			Food, drink and tobacco		Chemicals, coal and petroleum products		Metal manufacture		Engineering and allied industries				
										Engineering				Ship-building and marine engineering
	Coal mining	MLH 104[1]	Other mining and quarrying	Food	Drink and tobacco	Coal and petroleum products	Chemicals and allied industries	Ferrous	Non-ferrous	Total engineering	Mechanical engineering	Instrument engineering	Electrical engineering	
MLH[2]	101	104	102, 103 109	211-229	231-240	261-263	271-279	311-313	321-323	331-369	331-349	351-354	361-369	370
Weights	34	0·3	7	54	23	9	57	35	12	170	92	12	66	14
1973	103·2		121·1	106·0	96·3	120·6	105·3	129·5	117·0	98·4	97·1	92·4	101·2	95·4
1974	85·4		112·2	103·9	99·1	116·0	109·5	116·2	113·3	102·3	101·2	98·7	104·5	98·9
1975	100·0	100	100·0	100·0	100·0	100·0	100·0	100·0	100·0	100·0	100·0	100·0	100·0	100·0
1976	92·7	5 123	96·3	103·0	101·3	105·7	112·0	104·7	105·4	96·6	95·4	97·7	98·1	96·5
1977	90·1	15 023	95·8	105·2	101·0	102·7	115·9	103·1	108·3	97·5	94·0	100·1	102·0	93·5
1978	89·2	21 796	103·9	106·3	106·1	101·4	117·0	102·3	106·9	99·4	92·4	106·2	107·9	86·4
1979	89·3	31 333	106·0	107·3	108·4	105·3	119·3	104·4	105·3	101·2	91·3	109·4	113·6	78·1
1980	91·9	32 110	97·4	106·5	106·9	93·3	109·9	67·1	97·3	97·6	85·4	107·1	112·9	67·8
Seasonally adjusted														
1976														
1st quarter	94·8	2 400	98·4	100·0	95·5	102·3	107·7	100·3	101·2	96·1	96·7	95·3	95·3	96·9
2nd quarter	93·8	4 100	98·9	103·7	102·5	105·1	110·7	109·7	105·1	97·3	96·7	96·8	98·2	93·4
3rd quarter	91·1	5 300	94·1	104·5	105·6	105·3	113·4	106·1	106·1	96·0	94·2	96·5	98·4	92·7
4th quarter	91·3	8 600	93·8	103·8	101·7	110·1	116·4	102·6	109·2	97·2	94·1	102·2	100·6	103·0
1977														
1st quarter	91·2	12 800	100·0	106·8	101·1	105·6	117·9	106·8	113·1	98·8	96·9	98·7	101·4	92·1
2nd quarter	91·0	15 400	94·5	104·2	98·5	103·4	115·8	103·5	107·7	96·2	93·0	100·2	99·9	96·5
3rd quarter	89·7	15 500	94·6	105·1	100·4	101·6	116·0	105·1	105·7	97·6	93·6	100·9	102·6	95·9
4th quarter	88·4	16 500	94·1	104·7	104·0	100·2	113·9	97·0	106·9	97·6	92·6	100·6	104·1	89·2
1978														
1st quarter	89·5	18 500	95·1	104·4	105·3	98·9	114·4	107·1	105·1	98·2	92·4	102·3	105·5	89·5
2nd quarter	89·1	21 400	102·0	108·2	107·8	99·2	116·7	107·0	108·6	98·9	92·1	105·7	107·2	87·9
3rd quarter	89·4	22 200	105·8	107·7	103·3	104·6	118·8	97·0	107·1	100·6	94·0	106·5	108·8	85·7
4th quarter	88·7	25 000	112·8	104·6	108·1	102·9	118·3	98·0	106·7	99·8	91·1	110·2	110·2	82·6
1979														
1st quarter	88·0	29 300	96·4	105·9	105·2	101·2	114·9	98·5	105·0	99·8	89·8	107·1	112·4	82·4
2nd quarter	87·5	31 800	110·3	107·2	109·3	103·4	122·3	112·8	107·6	104·2	94·4	110·7	116·8	81·9
3rd quarter	90·8	33 000	110·3	108·3	109·7	109·3	121·0	103·2	103·8	97·6	87·6	108·2	109·7	75·9
4th quarter	91·0	31 200	107·0	107·8	109·4	107·2	118·7	103·0	104·8	103·3	93·6	111·5	115·5	72·1
1980														
1st quarter	92·4	32 600	107·8	106·9	114·7	98·8	119·2	41·4	104·7	103·7	93·0	110·4	117·6	70·3
2nd quarter	90·7	31 700	99·7	105·5	104·4	93·4	110·1	91·5	102·5	98·2	86·8	107·8	112·3	68·1
3rd quarter	91·4	30 700	93·9	105·8	103·7	90·3	105·4	70·3	95·1	97·1	83·7	109·0	113·6	66·7
4th quarter	93·0	33 500	88·2	107·9	104·9	90·7	104·7	65·1	87·0	91·5	78·2	101·4	108·2	66·1
1981														
1st quarter	90·4	35 000	88·2	106·2	108·7	86·0	105·7	73·1	86·2	89·7	78·4	98·7	103·8	65·1
2nd quarter	90·7	34 500	90·1	104·5	96·1	83·4	108·7	77·4	84·7	89·2	76·3	97·3	105·8	66·0
1980														
August	90		94	105	101	90	102	69	100	97	84	109	114	66
September	92		94	107	107	86	106	66	88	96	82	108	113	67
October	93		88	109	107	89	106	63	89	93	79	103	110	67
November	93		88	106	105	91	102	65	94	91	78	101	108	66
December	93		89	108	103	91	106	68	78	90	77	99	106	66
1981														
January	92		88	107	110	89	104	69	87	90	78	99	104	66
February	89		88	106	117	85	107	74	87	90	78	98	103	65
March	90		88	106	99	84	106	76	84	90	78	98	104	65
April	90		89	105	98	89	107	72	86	89	77	97	105	66
May	91		92	105	97	82	105	75	88	89	76	97	106	65
June	91		89	104	93	78	115	85	81	89	76	98	107	67
July	90		90	103	100	81	112	74	85	90	76	100	107	67
August	89		89	107	104	82	114	75	89	90	..	..		68

1. MLH 104: extraction of mineral oil and natural gas.
2. Industries are grouped according to the Standard Industrial Classification 1968.
.. Not available.

Source: Central Statistical Office

Table 5.6 Pay of Astrologers

Year	1975 (Nov)		1976 (Nov)		1977 (Nov)		1978 (Nov)		1979 (Nov)
Basic pay (£ per year) / Stage of career	Actual pay	Equivalent at 1979 rates	Actual pay at	Equivalent 1979 rates	Actual pay	Equivalent at 1979 rates	Actual pay	Equivalent at 1979 rates	Actual pay
On appointment	2610	4359	2813	4245	3015	4131	3683	4420	4434
After 1 year	2727	4554	2939	4435	3150	4316	3848	4618	4635
After 3 years	2997	5005	3232	4874	3467	4743	4229	5075	5094
Long serving	3411	5793	3675	5595	3939	5396	4811	5773	5793

Table 5.6 is an example of a good 'amateur' effort to present figures for each year from 1975 to 1979 on the pay of people at various specified stages of a career in a professional group. The idea is to show how actual pay, and its equivalent at 1979 pay rates (adjusted for inflation) varied over the years.

The purpose of this example is to show how even a good 'amateur' table can be greatly improved by applying the 'professional' approach recommended in this chapter. You will see many worse tables than Table 5.6 but by our criteria it has a number of faults:

—inadequate title

—geographical coverage not stated

—too many ruled lines

—obscure indication of what the figures in the table are (basic pay in £ per year)

—variation of pay over time is not shown clearly because years have been arranged horizontally instead of vertically

—the numbers have no commas or spaces following the thousands digits.

Table 5.7 is a 'professional' version of Table 5.6. As well as providing easier reference to the figures it also serves a demonstration purpose by showing almost at a glance the main feature of the data. This is that, taking inflation into account, pay at each career stage fell slightly between 1975 and 1977 but in 1978 jumped back to the 1975 level and remained at this level in 1979.

Table 5.7: Basic pay of Astrologers 1975 to 1979 and equivalent at 1979 rates

United Kingdom   *£ per year*

Stage of career	Year (Nov)	Actual pay	Equivalent at 1979 prices
On appointment	1975	2,610	4,360
	1976	2,810	4,250
	1977	3,020	4,130
	1978	3,680	4,420
	1979	4,430	4,430
After one year	1975	2,730	4,550
	1976	2,940	4,440
	1977	3,150	4,320
	1978	3,850	4,620
	1979	4,640	4,640
After 3 years	1975	3,000	5,010
	1976	3,230	4,870
	1977	3,470	4,740
	1978	4,230	5,080
	1979	5,090	5,090
Long serving	1975	3,410	5,790
	1976	3,680	5,600
	1977	3,940	5,400
	1978	4,810	5,170
	1979	5,790	5,790

Table 5.8 Extract from *British Business*, 23 July 1982

Production of man-made fibres in March

In the three months ended March 1982 the volume of production of man-made fibres, seasonally adjusted was estimated to have been 4 per cent lower than in the three months ended December 1981. Total production by weight of man-made fibres in March 1982 was 15 per cent lower than in March 1981; output of continuous filament was 25 per cent lower and staple fibre 9 per cent lower than March 1981.

Inquiries: 01–211 7052 or 01–211 4673.

Index of the volume of production

1975 = 100

1973	129	1976	110	1979	102
1974	111	1977	98	1980	76
1975	100	1978	105	1981	65

	Actual	Seasonally adjusted		Actual	Seasonally adjusted
1980					
1st qtr	92	90	Sep	64	69
2nd	84	80	Oct	68	65
3rd	59	65	Nov	70	68
4th	70	69	Dec	72	74
1981			**1981**		
1st qtr	71	69	Jan	65	69
2nd	68	65	Feb	74	69
3rd	57	63	Mar	73	69
4th	62	61	Apr	74	69
			May	67	62
1982			June	64	65
1st qtr	61	59	July	56	56
			Aug	48	63
1980			Sept	68	72
Jan	98	103	Oct	64	62
Feb	90	85	Nov	63	61
Mar	87	82	Dec	60	60
April	98	93			
May	83	76	**1982**		
June	71	72	Jan	55	58
July	69	70	Feb	66	61
Aug	42	55	Mar	60	57

Man-made fibre production

Thousand metric tonnes

	Total	Continuous filament (singles)	Staple fibre
1975	562.5	246.5	316.0
1976	618.4	270.7	347.7
1977	551.8	239.3	312.5
1978	607.2	240.6	366.6
1979	596.3	230.6	365.7
1980	449.8	162.6	287.2
1981	394.6	126.3	268.3
1980			
1st qtr	133.9	51.9	82.0
2nd	125.5	42.4	83.1
3rd	89.0	30.9	58.1
4th	101.4	37.4	64.0
1981			
1st qtr	101.9	38.1	63.8
2nd	102.0	34.2	67.8
3rd	91.6	26.0	65.6
4th	99.1	28.0	71.1
1982			
1st qtr	92.0	28.5	63.5
1980			
January	49.5	19.1	30.4
February	40.5	16.8	23.7
March	44.0	16.0	28.0
April	48.2	16.2	32.0
May	41.8	14.6	27.2
June	35.5	11.6	23.9
July	36.0	11.6	24.4
August	21.2	7.9	13.3
September	31.8	11.4	20.4
October	33.9	12.0	21.9
November	34.0	11.6	22.4
December	33.6	13.8	19.8
1981			
January	32.3	12.0	20.3
February	32.7	12.8	19.9
March	37.0	13.3	23.7
April	36.3	12.4	23.9
May	33.5	11.7	21.8
June	32.3	10.2	22.1
July	30.3	8.9	21.4
August	25.2	7.2	18.0
September	36.1	9.9	26.2
October	33.7	9.9	23.8
November	32.3	9.6	22.7
December	33.2	8.6	24.6
1982			
January	29.6	8.2	21.4
February	30.9	10.3	20.6
March	31.5	10.1	21.4

Source: Production stats: Man-made Fibres Producers Committee. Index of production: Department of Industry.

Table 5.8 is reproduced from *British Business*, of 23 July 1982. The layout of this pair of tables was clearly affected by the need to save space but a number of improvements might have been made without using significantly more column inches. First, the lefthand table might have been laid out in the same way as the righthand one, which would make it easier to identify annual, quarterly and monthly figures; secondly these three sub-headings might appear in the tables and thirdly the column spacing of the righthand table might be altered to show a larger gap between the total column and the other two: this can be achieved either by abbreviating the long column heading 'Continuous filament (singles)' or by putting the total column on the right and showing staple fibres first (see Table 5.9, overleaf).

Table 5.9 Production of man-made fibres, March 1982

☑

Index of the volume of production		Man-made fibre production		
1975 = 100				Thousand metric tonnes

Annual data

	Index		Staple fibre	Continuous filament (singles)	Total
1973	129				
1974	111				
1975	100				
1976	110	1975	316.0	246.5	562.5
1977	98	1976	347.7	270.7	618.4
1978	105	1977	312.5	239.3	551.8
1979	102	1978	366.6	240.6	607.2
1980	76	1979	365.7	230.6	596.3
1981	65	1980	287.2	162.6	449.8
		1981	268.3	126.3	394.6

Quarterly data

	Actual	Seasonally adjusted		Staple fibre	Continuous filament (singles)	Total
1980			**1980**			
1st qtr	92	90	1st qtr	82.0	51.9	133.9
2nd	84	80	2nd	83.1	42.4	125.5
3rd	59	65	3rd	58.1	30.9	89.0
4th	70	69	4th	64.0	37.4	101.4
1981			**1981**			
1st qtr	71	69	1st qtr	63.8	38.1	101.9
2nd	68	65	2nd	67.8	34.2	102.0
3rd	57	63	3rd	65.6	28.0	91.6
4th	62	61	4th	71.1	28.0	99.1
1982			**1982**			
1st qtr	61	59	1st qtr	63.5	28.5	92.0

Monthly data

	Actual	Seasonally adjusted		Staple fibre	Continuous filament (singles)	Total
1980			**1980**			
Jan	98	103	Jan	30.4	19.1	49.5
Feb	90	85	Feb	23.7	16.8	40.5
Mar	87	82	Mar	28.0	16.0	44.0
April	98	93	April	32.0	16.2	48.2
May	83	76	May	27.2	14.6	41.8
June	71	72	June	23.9	11.6	35.5
July	69	70	July	24.4	11.6	36.0
Aug	42	55	Aug	13.3	7.9	21.2
Sept	64	69	Sept	20.4	11.4	31.8
Oct	68	65	Oct	21.9	12.0	33.9
Nov	70	68	Nov	22.4	11.6	34.0
Dec	72	74	Dec	19.8	13.8	33.6
1981			**1981**			
Jan	65	69	Jan	20.3	12.0	32.3
Feb	74	69	Feb	19.9	12.8	32.7
Mar	73	69	Mar	23.7	13.3	37.0
April	74	69	April	23.9	12.4	36.3
May	67	62	May	21.8	11.7	33.5
June	64	65	June	22.1	10.2	32.3
July	56	56	July	21.4	8.9	30.3
Aug	48	63	Aug	18.0	7.2	25.2
Sept	68	72	Sept	26.2	9.9	36.1
Oct	64	62	Oct	23.8	9.9	33.7
Nov	63	61	Nov	22.7	9.6	32.3
Dec	60	60	Dec	24.6	8.6	33.2
1982			**1982**			
Jan	55	58	Jan	21.4	8.2	29.6
Feb	66	61	Feb	20.6	10.3	30.9
Mar	60	57	Mar	21.4	10.1	31.5

Source: Production stats: Man-made Fibres Producers Committee. Index of production: Department of Industry.

Laid out like this, the commentary would appear across the top or bottom of both tables.

The accompanying verbal summary to Table 5.8 is criticised in Chapter 8 on page 95. It can certainly be improved. But note that a telephone number for further enquiries is prominently displayed and the source of the data is quoted below the table.

Chapter 6: Charts

6.1 Do charts work?

Charts are popular with both authors and readers of reports and technical articles. But they often fail as tools of communication. It is useful to probe this paradox by discussing the reasons for both their popularity and their frequent failure.

First, why are they popular with those who construct them? There seem to be two reasons. One reason is that the originator of a chart usually expects it to please the reader. The other, and more important, reason is that once a set of data has been studied and the main pattern in the data clearly identified, a well planned chart (or set of charts) is often the best way of describing these patterns concisely and memorably. Sometimes a picture may indeed be worth a thousand words.

Does it please the reader? Pause for a moment and consider your own reaction on turning a page and finding a clear, apparently simple graph on the next page. Many people will experience two reactions: first relief that there is less than a full page of text to be read and secondly relief that some part of the argument has been condensed into a simple picture which can be scanned quickly. So a chart *is* likely to please the reader. But is it likely to do its job of communicating effectively? Here the answer is much more doubtful.

Remember the communicator's state of mind when he produced the chart? He had studied and assimilated the data and identified the main patterns. He was, in fact, thoroughly familiar with:

—what the data represented

—the units of measurement

—the geographical coverage

—the time period covered

—the range of variation in the data.

All these facts were already part of his store of knowledge and the chart represented, *for him*, a concise and efficient way of arranging the data in order to highlight the salient features against this factual background.

Consider, for example, Figure 6.1 (overleaf) from the 1983 edition of *Social Trends*.

The originator of this chart was completely familiar with the definition of real household disposable income, for which the reader is invited to consult part 5 of the Appendix to *Social Trends* 13. The definition found there makes it sound rather complex:

'Real household disposable income
Household disposable income is equal to the total current income of the household sector less payments of United Kingdom taxes on income, employees' national insurance contributions and contributions of employees to occupational pension schemes. It is revalued at constant prices by a deflator implied by estimates of total household expenditure at current and constant prices. This deflator is a modified form of the consumers' expenditure deflator.'

but essentially it is the amount of money actually available for spending, measured in terms of its buying power. The originator had also decided to show graphically the *percentage change over the preceding year* and to display this measurement for the 10-year period from 1971 to 1981. He or she knew that the largest positive change between successive years was something under 10 per cent and that the biggest decrease was just over 2 per cent. In fact the writer clearly knew a great deal more about these data including the original numbers from which the annual changes were calculated. The data are actually presented to the reader in a table on the page before the chart. The relevant lines from this table and the text accompanying the table and chart are reproduced overleaf.

Figure 6.1　Annual changes in real household disposable income per head

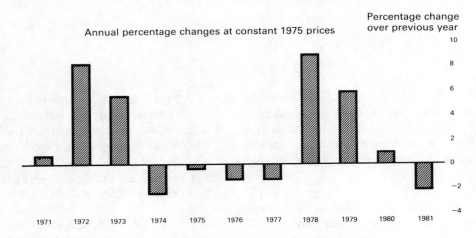

Annual percentage changes at constant 1975 prices

Percentage change over previous year

Verbal summary to Table 6.1

Total household disposable income increased from £36 billion in 1971 to £159 billion in 1981. In real terms (ie allowing for inflation) this was an increase of 24 per cent. Real household disposable income per head increased by a similar percentage over the same period although, as shown by [Chart 5.2], the annual changes from one year to the next varied widely, with small reductions in the mid-1970s and again in 1981.

The person who planned the chart has devised an effective way of highlighting the change from 90 to 97 between 1971 and 1972 and the even greater increase from 97 to 106 between 1977 and 1978. By contrast with these years of increase, we see small year-on-year reductions from 1974 to 1977 and again in 1981.

But what about the reader? How long has it taken you to read the last few paragraphs (from Figure 6.1)? And are you sure that you understand exactly what is the main message of Figure 6.1? Without looking back at the figure or the table, can you describe the main pattern in the data using your own words? And can you say whether or not real household expenditure per head was greater in 1981 than in 1978?

Referring to the bottom line of the table we see that it was in fact greater, 111 as opposed to 106. But to retrieve this information from the graph the reader would have to note that the small decrease of about 2 per cent between 1980 and 1981 was preceded by year-on-year increases of

about 6 per cent in 1979 and about 1 per cent in 1980. The net effect must therefore be that the 1981 value is *higher* than the 1978 one.

This is not meant to suggest that the originator of the graph intended the reader to use it to answer such a detailed question. The question is merely posed to emphasise the typical difference between the state of knowledge of those who plan charts and those who read them.

What has just been said illustrates the main limitation on the use of graphs. A graph will communicate its message effectively only to someone who has mastered the necessary background information about units of measurement, the scale used and range covered on each axis as well as any conventions used on the graph (for example, the use of a dotted line to represent, say, the UK figure and a solid line to represent the corresponding figure for a specific region). If you are confident that your reader is familiar with all these details you may then safely rely on a well executed graph to communicate a simple message: you may even choose to use quite a complicated form of graph *so long as both you and your reader are skilled in interpreting its conventions*. But beware of underestimating how long it takes to build up such skill and, if in doubt, assume that few of your audience will be familiar with either the background to the data or any but the most common graphical conventions.

A second problem with using graphs is that most

Table 6.1　Household income[1]: national totals

United Kingdom

	1971	1972	1973	1974	1975	1976	1977	1978	1979	1980	1981
Total household disposable income (£s thousand million)	36.0	41.5	47.8	54.9	67.7	77.2	87.7	103.0	124.0	146.9	159.4
Real household disposable income per head (index numbers—1975 = 100)	90	97	103	100	100	99	97	106	112	113	111

[1] See Appendix, Part 5: The household sector.

Source: Central Statistical Office

people have not been trained in how to *read* them rather than merely look at them. A graph looks like a picture and it is well known that we have only to glance at a picture for a second or two in order to recognise it when we meet it again. This encourages us to register only the obvious pattern in a graph before passing back to the text.

On several courses at the Civil Service College a film is regularly shown which includes the following graph:

Figure 6.2 Average weekly cinema admissions

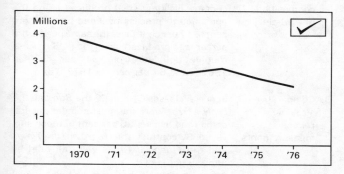

along with the favourable commentary 'the story of the cinema in Britain in a single snapshot'.

During the discussion period after the film the audience is asked what they remember about that particular chart. Almost everyone can trace the outline curve in the air with

comments like 'down, slightly up, then down again'. But when asked what time period was covered by the data and what were the units on the vertical axis, very few people can volunteer answers. Those who can are almost invariably either expert users of graphs or else have some particular interest in cinema attendances. The first group has got into the habit of reading both axes when they look

at a graph and remembering the units of measurement when they study its shape: the second group already has some background information about cinema admissions and the graph represents either data with which they are already familiar or data which can be related to their existing store of knowledge, but only by also absorbing the detailed information on the axes. The real non-specialists in the audience have mostly only absorbed the striking outline of the graph before turning their attention to the next point.

So where does this leave us? Can charts ever be relied on as effective communication tools for a non-specialist audience? Yes indeed; they can work very well provided one recognises the limitations on the sort of message which can be transmitted graphically and gives due thought to the choice of chart and how it should be presented.

Charts can work extremely well for showing relationships in the data—features like:

—the comparative sizes of the parts which make up a total

—a dramatic change in one contributor's share of a total

—the comparative sizes of a number of related measures

—a striking pattern of growth or decline over time (in one or possibly several related series)

—a change in the ranked order of a set of related measurements

—a correlation between two sets of data.

Patterns like these can be effectively encapsulated in a well chosen chart, but detailed numerical information cannot. For example, we see immediately from Figure 6.3 that the percentage of liquid steel produced using the Basic Oxygen Steelmaking (BOS) process approximately doubled between 1968 and 1978, while the percentage produced by the open hearth process declined

Figure 6.3 Liquid steel production by process: British Steel Corporation, 1967-68 and 1977-78

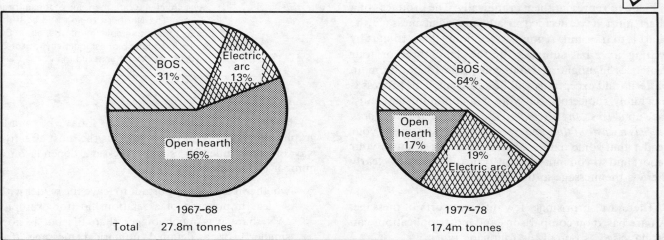

dramatically. These are the messages which the eye registers; not the numbers 31%, 64%, 56%, 17% (even although these figures appear on the chart).

If it is important that the reader remembers these numbers, the chart alone is unlikely to be enough.

In fact, this pie chart is unlikely to serve any useful purpose in a report unless it is accompanied by a verbal summary—something like:

Between 1967/68 and 1977/78 the percentage of liquid steel produced using the open hearth process fell sharply from 56 per cent to 17 per cent. Over the same period, the percentage produced using the BOS process roughly doubled, from 31 per cent in 1967/68 to 64 per cent in 1977/78.

Quoting the actual figures in the summary performs two useful functions. Firstly it confirms to the reader that he has interpreted the chart correctly and secondly it links the numbers with the corresponding pie segments in the reader's mind: there is now a fair chance that the whole message will be stored in his memory.

This example illustrates the two basic rules for producing an effective chart:

1. Keep it simple

2. Talk the reader through it.

6.2 When to use a chart

Graphs are time consuming and expensive to produce: they must therefore be used economically. It has been said that graphs should be used:

—to reveal the unexpected

—to make the complex easier to perceive.

To this should be added, for the purpose of communicating to a non-specialist audience

—to convey concisely and memorably a message (which may or may not be unexpected).

Just as in the case of a table, the data must be studied and the salient features decided upon before any attempt is made to represent them graphically. The producer of a chart must have decided in advance what message the chart is to transmit. A good way of doing this is to start by writing a verbal summary of the data: three or four sentences highlighting what you judge to be the main patterns and exceptions will be sufficient. This process is inevitably subjective, but you as a communicator must face up to this. *Any attempt to use a chart as a neutral data store is a waste of time and effort.* You have promised your reader that some data are relevant at this point in your report and so you must demonstrate this by saying clearly 'here is the message in the data'.

There are surprisingly few different sorts of message. All are based on comparisons and most applications can be classified as one of the following types:

Table 6.2 Comparisons and typical summaries

Sort of comparison	Typical summary
1. Components as part of a whole	In January 1981, 50 per cent of 16–18 year olds were in employment and 29 per cent were in education. 15 per cent of this age group was unemployed and the remaining 6 per cent were on Youth Opportunities Programmes.
2. Change in composition of a whole	Between 1967/68 and 1977/78 the percentage of liquid steel produced using the open hearth process declined from 56 per cent to 17 per cent. Over the same period, the percentage produced using the BOS process roughly doubled, from 31 per cent in 1967/68 to 64 per cent in 1977/78.
3. Comparative sizes of related measurements	In social classes C1 and C2 the *Sun* and the *Daily Mirror* were the most popular papers, being read by over 30 per cent of those surveyed. By contrast, the *Guardian*, the *Times* and the *Financial Times* were all read by under 5 per cent. Approximately 28 per cent of those social classes read no daily paper.
4. Change over time of one or more related measurements	From 1970 until 1982 the number of vacancies notified to unemployment offices fluctuated within fairly narrow limits between 100 thousand and 400 thousand. Over the same period the number unemployed changed dramatically, climbing virtually continuously from 1974 when there were 500 thousand unemployed until 1982 when the number of unemployed reached 2.8 million.
5. Change in the ranked order of a set of related measurements	In 1971 almost 40 per cent of secondary school pupils attended secondary modern schools, 34 per cent attended comprehensives and 18 per cent attended grammar schools. By 1978, the pattern had changed noticeably with 77 per cent of secondary pupils in the maintained sector attending comprehensives, 10 per cent attending secondary modern schools and only 5 per cent attending grammar schools.
6. Relationship between two sets of data	In general, areas with high unemployment also had high inactivity rates for 60–64 year old men although Scotland, with a higher unemployment rate than Wales, the Northwest or Yorks and Humberside, recorded a slightly lower inactivity rate than any of these three regions. The Northern region had both the highest male unemployment rate, about 8.5 per cent, and the highest inactivity rate of 21 per cent for 60–64 year old men.

All these sorts of comparisons may be made graphically: whether or not a graph is the most effective way of communicating such a message depends on a number of factors including:

—whether it is more important to leave the reader with a clear impression of a relationship (for example, 'A's share was much bigger than B's but is now smaller') or a detailed numerical message (for example, 'dividends increased from 4p to 7p per share' or 'education accounted for 72 per cent of total expenditure'): generally speaking, graphs convey relationships better than tables or text;

—whether or not the message to be transmitted lends itself to a clear graphical statement (for example, if two series have grown in parallel over a 20-year period this can be shown concisely using a set of line graphs; conversely, four series having dissimilar patterns of growth or decline over a five-year period are likely to produce a confused tangle of lines);

—the scales of measurement (for example, can the range of both variables be shown comfortably on the same scale?);

—the skill with which the chart is planned and executed (skill in planning comes with practice and a continual critical appraisal of all charts, your own and other people's: skill in execution usually requires professional help);

—the reader: there is not much you can do about this unless, of course, the report is aimed at a small specific audience in which case it pays to get to know their idiosyncrasies; some people have graphic minds and others have not, but for both sorts of reader a good chart is more likely to prove successful than a poor one.

6.3 What sort of chart?

Assuming that you have decided to use a chart to illustrate a particular point, what sort of chart should be used? For a non-specialist audience it is probably safest to choose from a limited range of three or four basic varieties:

pie charts

bar charts (in a wide variety of forms)

line graphs

and, perhaps, scatter diagrams.

The reason for this is that most people will have no difficulty in interpreting these sorts of chart. If you use a more complicated or unusual diagram there is a danger that the reader will not have the time or inclination to interpret the conventions used in the chart and will fail to assimilate the message.

A helpful booklet which matches up the kind of comparison to be illustrated with possible forms of chart is *Choosing and Using Charts* by Gene Zelazny.[1] The following table borrows heavily from that booklet.

[1] Published by Video Arts, New York, 1972.

Table 6.3 Which chart for which comparison?

Sort of comparison	Possible charts
1. Components as part of a whole	Pie chart. Component bar chart.
2. Change in composition of a total	Pie chart. Component bar charts (vertical or horizontal bars).
3. Sizes of related measurements	Grouped bar charts (vertical or horizontal bars). Possibly isotypes.
4. Change over time in one or more related measurements	Line graph. Column bar charts.
5. Change in ranked order of a set of measurements	Paired or grouped bar charts.
6. Relationship between two sets of measurements	Back-to-back bar charts. Scatter diagrams.

6.4 How to draw charts.
General guidelines

If a chart is to communicate effectively, the reader must find it satisfying to look at and easy to understand. The following guidelines apply to all charts:

1. Make sure your reader is in no doubt about:

—the kind of objects, events, people or measurements represented

—the units used

—the geographical coverage

—the time period covered

—the scale of measurements

—the source of the data

—how to interpret the chart.

All this information should appear round the edges of the chart: that is, in the headings, the labels on the axes and, if necessary, in a clear key (but see 2. below on the subject of direct labelling).

2. Make it easy to read. This means:

—make it no bigger than is necessary for clarity: it is much harder to assimilate information from a large chart than from a small one; A4-size is usually much too big; this can be demonstrated by giving someone an A4-sized chart to look at; the first thing he will do is to hold it at arm's length to reduce its effective size;

—keep it the right way up: it is distracting to have

to rotate a document through 90 degrees in order to read it; keep all the ancillary markings like axis labels the right way up, too;

— label sections of pie charts, component bar charts and line graphs directly rather than by using an explanatory key at the bottom or side of the chart: continual reference to a key interferes with the reader's prime task of assimilating the main features of the chart;

— include important numbers (for example, the percentages represented by key slices) on the chart if this can be done without loss of visual clarity;

— use clearly differentiated shadings, colours or symbols to discriminate between different segments of pies or component bar charts and between different lines.

3. Make it easy to understand. This means:

— the conventions used in constructing the graph should be easy to interpret (for example, to represent quantities of different magnitudes use bars of equal widths but varying lengths: the non-specialist reader should not be asked to interpret both changes in length and changes in width);

— each chart must be accompanied by a clear verbal summary of its salient features. The verbal summary should be as close as possible to the chart and direct the reader to the chart (as in 'Figure X, below' or 'Figure Y, opposite'). It is helpful to quote directly the important numbers on the chart in the verbal summary: this helps to link the chart and the numbers in the reader's memory. But the summary should only comment on a few important points, it should not be a blow-by-blow account of each feature;

— do not try to include too much information on a single chart: the reader will find it easier to follow a story built up using a series of simple charts than to disentangle a single complicated diagram.

The following chart and commentary, taken from *Social Trends* 10, illustrates many of these points.

The commentary which appears opposite the chart says 'There are differences in marriage and family building patterns between the New Commonwealth and Pakistan (NCWP) born and UK born populations, and between different ethnic groups themselves. Women born in the West Indies are far more likely to have children born outside marriage as compared to women born in the United Kingdom. But women born in the Indian subcontinent are far less likely to have children born outside marriage than women born in this country.'

Minor improvements might be suggested (for example, some of the percentages printed on the chart might have been quoted in the verbal summary) but note the following virtues:

1. The information needed to understand it is all provided clearly on the chart:

we are dealing with *births in 1977*

they are analysed by *birthplace of mother* (ie West Indies, United Kingdom or India, Pakistan or Bangladesh)

the data were provided by the Office of Population Censuses and Surveys.

Figure 6.4   Births: by birthplace of mother, Great Britain, 1977

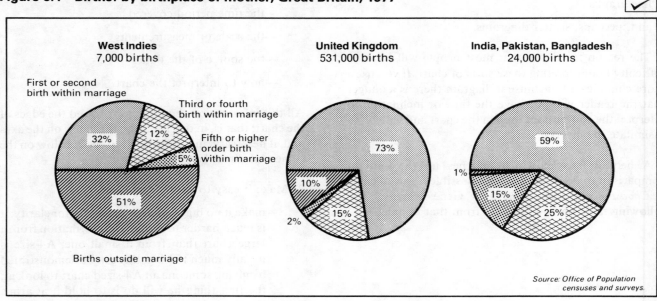

Source: Office of Population censuses and surveys.

2. It is easy to read—that is:

—it is compact and the right way round (no need to twist the book to read the chart)

—the interpretation of each shading is shown by labelling the slices of the first pie directly ('Births outside marriage', 'First or second birth within marriage', etc.)

—the percentages represented by each slice appear on the pie

—the slices of pie are differentiated by shadings which are not likely to be confused visually.

3. It is easy to understand—that is:

—the conventions used in constructing the chart are easy to interpret: each pie is the same size and each is divided into four slices whose magnitudes represent the percentage of births falling into four defined categories (the fact that each pie represents different total number of births can easily be read from the labels above each pie);

—the verbal summary concentrates on a single striking feature, namely the difference in the percentage of births outside marriage between mothers from the West Indies, the UK and the Indian subcontinent.

This chart was drawn by a professional, a practice strongly recommended whenever possible. Unfortunately not all professional draughtsmen and illustrators know how to draw good statistical charts. If yours does not, a good idea is to supply a freehand sketch, giving layout and proportions. If you decide to draw your own charts you will have to learn some of the professionals' tricks. Useful publications on this topic are Schmids' books on graphic presentation and the ANSI guidelines on *Illustrations for Publication and Projection* (see Appendix D for details).

The three basic principles (charts should be clearly and completely labelled, they should be easy to look at, and easy to understand) apply to all charts. Their application to the four types of chart recommended for non-specialist audiences (pie charts, bar charts, line graphs and scatter plots) is discussed in Chapter 7.

6.5 Chatty charts

Charts are often used to good effect by journalists. In publications like the *Economist* or in the business section of quality newspapers, quantitative information is frequently displayed graphically. Here the design of charts and tables tend to be informal: catchy captions are often used and the basic chart may be enlivened in order to amuse (and hence interest) the reader. See Figure 6.5 below (from the *Economist* January 22, 1983).

When writing for the general public, such tricks are extremely useful and may certainly be used so long as:

—the data are displayed clearly and honestly

—your graphics designer is sufficiently talented to execute the design well.

Figure 6.5

Cash will do nicely, sir

Remember how plastic money was going to produce the cashless society? It hasn't happened. In the United States, notes and coins still add up to $715 for every American citizen over the age of 15. The British make do with pounds and pence worth $425; West Germans with $690. The fattest wallets are in Switzerland, where $2,020 worth of Swiss francs circulates for every adult. Much of the cash floats in companies and shop tills, though plenty must be under mattresses.

In most countries, credit cards and the increasing use of bank accounts have made cash less popular. Yet notes and coin now loom larger in the narrow definition of money supply, M1, in America and Italy than they did in 1970 (see table). And cash still represents 4.2% of gross domestic product in the United States (compared with 5% in 1970), 5.3% in West Germany (versus 5.4% in 1970), and 12.3% in Switzerland (13.2% in 1970).

The Swiss are as conservative about cash as about everything else: they rarely use credit cards for daily transactions. Foreign savers—especially migrant workers and the *Grenzarbeiter* (workers who commute across the Swiss border daily) are as keen as ever on owning bundles of Swiss notes. This has affected the D-mark, too.

Demand for D-marks also has other quirks. Officials at West Germany's Bundesbank say that the amount of D-mark cash in circulation can, for instance, vary with the stability of the Turkish government; doubts about Turkey's future make migrant workers' families extremely wary of changing their remittances into Turkish currency.

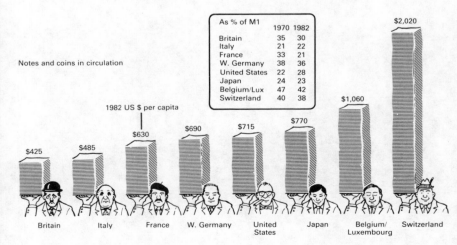

As % of M1	1970	1982
Britain	35	30
Italy	21	22
France	33	21
W. Germany	38	36
United States	22	28
Japan	24	23
Belgium/Lux	47	42
Switzerland	40	38

6.6 Special charts

This category is taken to include charts used by specialists (among themselves) and special charts designed to communicate with a non-specialist audience.

The three basic principles described in section 6.4 still apply to these graphs; the only difference is that a greater, more specialised store of background knowledge is assumed to be shared by the producer of the chart and the reader. Specifically, the reader is assumed to be familiar with more complex graphical conventions than the simple use of heights or areas to represent relative magnitudes.

Common examples of graphical conventions familiar to specialists are:

—the use of logarithmic scales by economists;

—cumulative sum, or 'cusum' charts, in which the variable of interest is represented by the slope of the graph rather than its height; such charts are very good for detecting systematic changes in processes with a strong random component;

—scatter plots of residuals, used by statisticians to assess the goodness of fit of a regression model.

Among specialists, writing for specialists, such techniques may be used safely and are often far more effective for particular purposes than conventional bar charts, pie charts or line graphs. Moreover the specialists we refer to here are not necessarily statisticians or economists but could be, for example, accountants, engineers or military officers who are expert in fields served by the statistics. Our reason for not dealing in more detail with specialist charts is not that we do not recognise their value but simply that their lack of universal applicability puts them outside the scope of this book.

Currently there is a vogue for using graphical methods (many of them recently devised) to explore data. Once the techniques have been mastered and their properties thoroughly understood there is no doubt that they can provide insights into data more easily than other methods. The danger is that in such circumstances, the same graphical technique may be used in presenting conclusions to an audience which has not mastered the properties of the unfamiliar technique. If this happens communication fails altogether.

There are of course circumstances in which the properties of a particular graphical device are so useful that it is worth teaching them to the audience before using this form of graph to present your conclusions. For example, you might wish to demonstrate that two series of data, of very different magnitudes, have grown at approximately the same rate over a 15-year period. This can be demonstrated elegantly and economically using semi-logarithmic graph paper—but only if the audience is aware that the use of a logarithmic scale has the effect of showing constant percentage growth as a straight line whose slope depends on the growth rate (and not on the absolute magnitudes plotted). To explain the properties of semi-logarithmic graph paper adequately to a non-specialist audience will probably require about half a page of text and a simple worked example. Whether this amount of effort, and distraction from the main report is justified by the effectiveness of the final graph can only be judged by the author. It may be; but it is important neither to underestimate how much effort is required to master the properties of an unfamiliar device, nor to underestimate the discouragement experienced by those who are presented with a chart which they cannot understand.

The process of educating the audience to understand an effective but unusual type of chart is much more likely to be worthwhile when you are producing a series of periodic reports to a limited (though not necessarily technical audience) than when you are writing a one-off report to a wide audience.

Chapter 7: Effective charts for general use

In Chapter 6 the following charts were recommended for communicating with non-specialists:

—pie charts

—bar charts (in a wide variety of forms)

—line graphs

—scatter charts

—isotypes

This chapter explains briefly how to draw each type of chart and gives examples of both good and bad practices.

All the charts are reproduced in black and white, even those which originally appeared in *Social Trends* where tones of red may have been used. The use of additional colour can certainly enhance the appearance of any chart and may make it easier to use; however the comments and criticisms of charts in this chapter are not dependent on the use of colour. It is always possible to design an effective *simple* chart in black and white and only simple charts are recommended in what follows.

7.1 Pie charts

Pie charts can be used effectively to illustrate:

a. the composition of a whole, that is, the relative sizes of different components;

b. the difference in composition of two or more related wholes.

A pie chart is constructed by dividing the 360 degrees at the centre of the circle into angles whose sizes are propor-

Figure 7.1 Activities of 16–18 year olds in January 1981

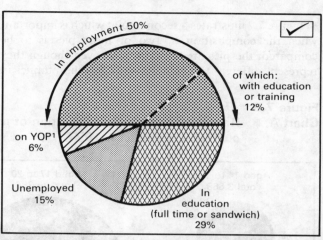

[1] Youth Opportunities Programme.

Source: Department of Education and Science, Scottish Education Department, Welsh Office, Department of Employment, MSC

tional to the sizes of the components represented. In the pie chart shown in Figure 7.1, 50 per cent of 16–18 year olds were in employment so the largest slice has an angle of 180 degrees (50 per cent of 360 degrees) at the centre of the pie; 29 per cent were in education so that slice of the pie has an angle of 104.4 degrees at the centre, and so on.

In order to make the composition of the pie as memorable as possible it is helpful to arrange the slices of a pie chart in some natural order. This will often be in decreasing order of size, or importance, with the largest slice positioned against the 9 o'clock line on the chart (or, if preferred, the 12 o'clock line). In Figure 7.1 the slices are shown in descending order of magnitude: in employ-

Figure 7.2 Liquid steel production by process: British Steel Corporation, 1967-68 and 1977-78

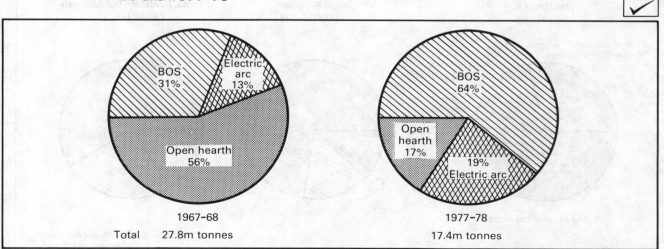

ment (50 per cent), in education (29 per cent), unemployed (15 per cent) and on YOPs (6 per cent).

Measuring from a common baseline is particularly helpful if the composition of two or more pies is to be compared. In Figure 7.2 (a copy of Figure 6.3) the most important feature is the increase in the percentage of liquid steel produced by the BOS process. The BOS slice is therefore positioned against the 9 o'clock line in each pie.

Figure 7.2 illustrates a second point which is important when the composition of two or more pies is to be compared: the pies are the same size, even though they represent different total amounts (27.8 million tonnes in the first case and 17.4 million tonnes in the second). Many statistical text books recommend that in cases like this the area of each pie should be proportional to the total represented. However, there is ample evidence[1] that most people are unable to estimate the relative areas of different circles accurately, and so carrying out the complicated calculations needed to scale the areas of two or three pies to represent varying totals correctly merely wastes time. Compare the two charts below, both taken from the same edition of *Social Trends*.

---

[1] For example: MEIHOFER, H J. The visual perception of the circle in thematic maps; experimental results. *Canadian Cartographer*. 1973, vol 10, pp 63–84.

Figure 7.3
Chart A   Remands to custody: proportion of males subsequently given custodial, and non-custodial, sentences, 1977

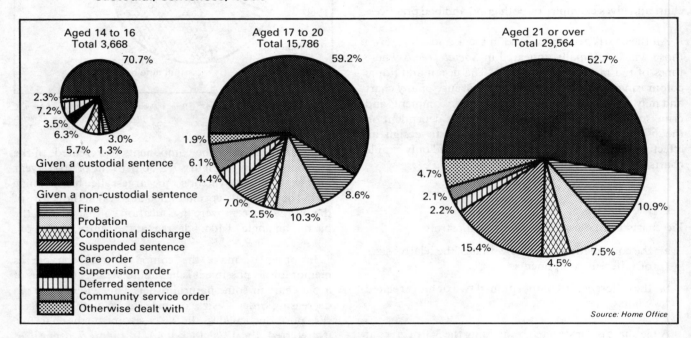

Chart B   Births: by birthplace of mother, Great Britain, 1977

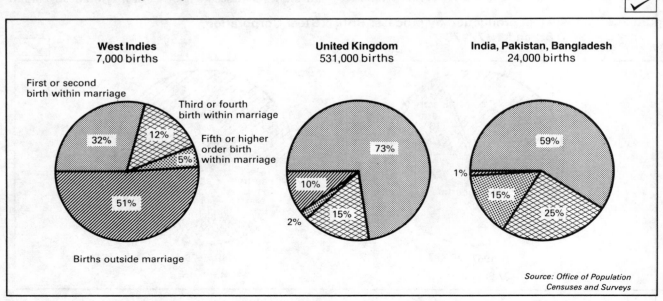

In chart A, the areas of the three pies have been scaled in the ratio of 3,668 to 15,786 to 29,564: but can you see, by looking at the circles, that the area of the middle one is about four times the area of the first and just over half the area of the third? Very few people can make such judgements accurately (and compare the percentage of custodial sentences given to each group as well). Most people will find Chart B easier to deal with. Here the relative sizes of different slices can be compared immediately (the reader being directed to the slices of greatest interest by the verbal summary, of course) and the fact that the vast majority of births were to mothers born in Britain is clearly recorded on the chart; 531 thousand births, compared with seven thousand births to West Indian mothers and 24 thousand to mothers born in the Indian subcontinent.

Figure 7.3 shows three pies on the same chart. This is probably the largest number of pies which should be presented on a single chart. Occasionally four pies can be effectively displayed in the same diagram but never more. To compare the composition of more than three totals, component bar charts should generally be used. (See section 7.2 page 72.)

It is counter-productive to include too many slices in a pie chart. You may expect your reader to show an intelligent interest in four or five categories but after that confusion and boredom will set in. The pie chart in Figure 7.4 does not work for this reason.

A possible exception to this is if you wish to convey a message like 'category A accounted for over 70 per cent of the total while the other 10 categories between them accounted for less than 30 per cent'. The pie chart in Figure 7.5 clearly conveys the message that just over half

Figure 7.4 Activities of 'Chemcorp' group

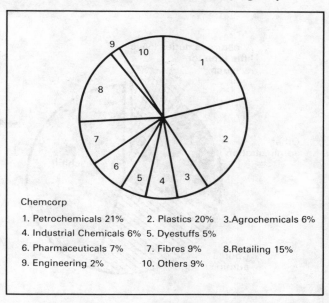

Chemcorp

1. Petrochemicals 21% 2. Plastics 20% 3. Agrochemicals 6%
4. Industrial Chemicals 6% 5. Dyestuffs 5%
6. Pharmaceuticals 7% 7. Fibres 9% 8. Retailing 15%
9. Engineering 2% 10. Others 9%

of the total group received custodial sentences. The details of the other categories are indistinct, and because of the number of categories shown, the producer of this chart has had to use a separate key to explain the shadings rather than label the chart directly.

Before moving on to discuss bar charts, it is worth mentioning two minor possible embellishments on pie charts. First the pie may be drawn as a symbol appropriate to the total whose composition is illustrated (so long as it is circular) for example a coin or a cake. Secondly, if a comparatively small slice of the pie is of particular interest, that slice may be shown lifted out of the rest of the pie as in the two examples in Figure 7.6.

Figure 7.5 Remands to custody: proportion of males aged 21 or over subsequently given custodial and non-custodial sentences

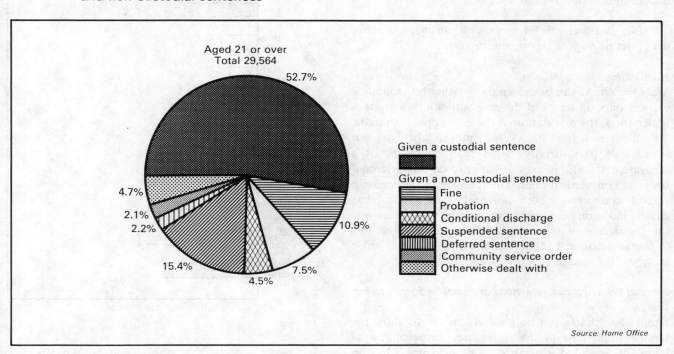

Aged 21 or over
Total 29,564

52.7%

Given a custodial sentence

Given a non-custodial sentence
Fine
Probation
Conditional discharge
Suspended sentence
Deferred sentence
Community service order
Otherwise dealt with

4.7%
2.1%
2.2%
15.4%
4.5%
7.5%
10.9%

Source: Home Office

71

Figure 7.6   Distribution of a college lecturer's time

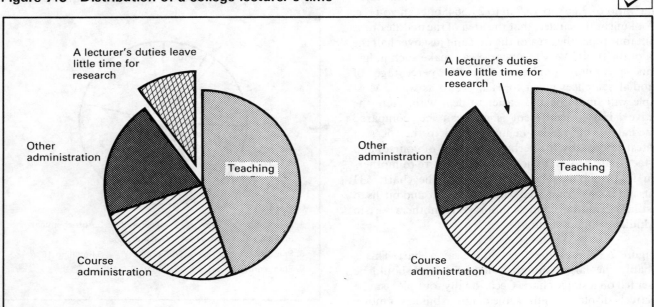

In Figure 7.6 the message being illustrated appears as a caption on the chart: 'A lecturer's duties leave little time for research'. This approach can be used to good effect whenever a chart illustrates a single clear message.

7.2   Bar charts

Bar charts are very versatile: they can be used to illustrate the magnitudes of related measurements, changes over time in one or more measurements, the composition of related totals, changes over time in the composition of a single total and also the degree of correlation between two measurements. A similar chart, the histogram, can be used to illustrate the frequencies with which different measurements occurred in a set of data.

Before discussing different uses of bar charts in more detail, let us dispose of one minor issue.

Horizontal or vertical bars?
Most bar charts can be drawn using either horizontal or vertical bars. In terms of the ease with which a chart is understood, the orientation of the bars appears to make little difference. It is, however, helpful to label bars and sections of bars directly rather than by means of a separate key and this may be easier to do on horizontal bars than on vertical ones. The following pages contain several examples of both horizontal and vertical bar charts, indicating that either sort can be used effectively (or ineffectively): the orientation of the bars is less important than the clarity with which the final chart matches its message.

Sizes of related measurements: bar charts and grouped bar charts
Simple bar charts can be used effectively to show the number of items in specified categories, the percentage of items in specified categories, or incidence rates in specified categories. In all cases bars of equal widths are drawn with lengths proportional to the measurement being illustrated.

Figure 7.7   Holidays taken abroad by United Kingdom residents in 1976

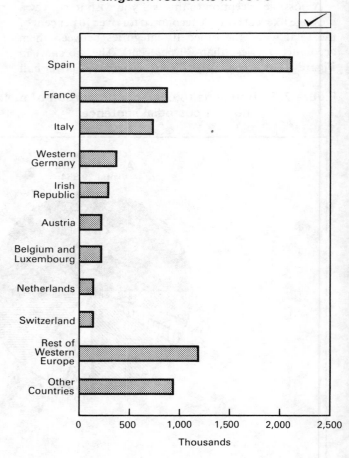

Adapted from Facts in Focus *1978, Chart 22*

Verbal summary to Figure 7.7

Figure 7.7 shows that 2.2 million UK holiday makers visited Spain in 1976. The next most popular single destination was France (0.9 million) and almost one million holidays were taken in foreign countries outside Western Europe.

Figure 7.8 shows the percentage of full-time university students studying different subject areas in 1975/76.

Figure 7.8 Percentage of university students studying different subject groups

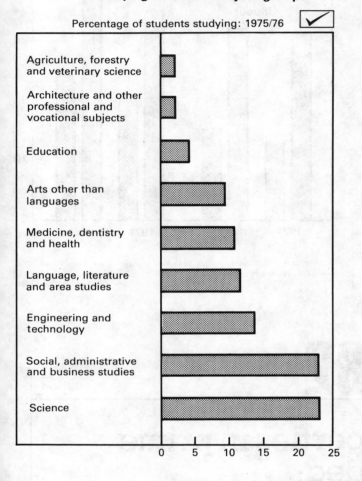

Percentage of students studying: 1975/76

Adapted from Facts in Focus *1978, Chart 15*

Verbal summary to Figure 7.8

The two most popular subject areas were science and the composite area social, administrative and business studies, each of which accounted for about 23 per cent of all full-time students.

Figure 7.9 shows the rate of alcohol misuse per 100,000 of adult population for both men and women of different ages in England in 1980.

Figure 7.9 Alcohol misuse—admissions[1] to mental illness hospitals and units: by sex and age, 1980

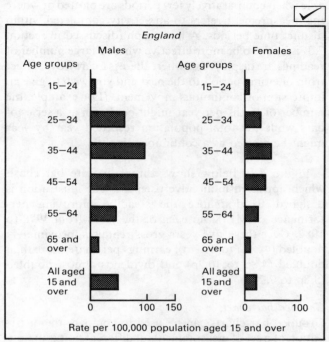

Rate per 100,000 population aged 15 and over

[1] All admissions with a primary diagnosis of alcoholic psychosis, alcohol dependence syndrome, or non-dependent abuse of alcohol.

Source: Department of Health and Social Security

Verbal summary to Figure 7.9

Alcohol misuse, as measured by admissions to mental illness hospitals, is roughly twice as prevalent among men as among women. The 'all ages' bar shows a male rate of about 60 per 100,000 population while the corresponding female rate is about 30 per 100,000. For both sexes, alcohol misuse is commonest in the age range 35 to 44.

In these three figures the bars are drawn horizontally, primarily for ease of labelling each bar. When using horizontal bars it is advisable to put the labels alongside and to the left of each bar if possible. If names of different lengths (for example 'Spain' and 'West Germany') are written to the right of or inside the bars, the eye no longer registers a clear profile of the relative lengths of each bar.

In Figure 7.9 the order of bars has been determined by the natural ordering of age groups. In Figures 7.7 and 7.8, where there is no natural order of categories, the bars have been arranged in order of size, the largest category at the top in one case and at the bottom in the other. Either arrangement can be used effectively.

Change over time: column bar charts
Change over time in a specified measurement can be illustrated using either a series of bars (usually drawn vertically) or as a line graph. A column bar chart is preferable when comparatively few periods are plotted or where the measurement refers to an activity completed within distinct time periods. A line graph (discussed in section 7.3) is likely to be more effective where a large number of readings are displayed or where there is a carry-over effect from one time period to the next and you wish to give an impression of continuous movement. (For example, the number of births per year might be shown as a series of bars while the total population recorded year by year might be plotted as a continuous line.)

Figure 7.10 below shows three separate bar charts which appeared in an advertisement. The illustration is designed to dramatise the steadily improving performance of MEPC in each of the years from 1978 to 1982. Over these five years gross rentals approximately doubled (44.1m to 88.9m); earnings per share more than doubled (4.8p to 10.2p) and dividends almost doubled (3.8p to 7.25p).

Grouped bar charts
Grouped bar charts can be used to compare the distribution of a set of measurements at different points in time or in different geographical areas. In these each bar is accompanied by one or more others showing the magnitude of the same measurement at a different time or in a different area. Pairs or triples of corresponding bars can

Figure 7.11 Participation[1] in indoor and outdoor sporting activities: by sex, 1977 and 1980

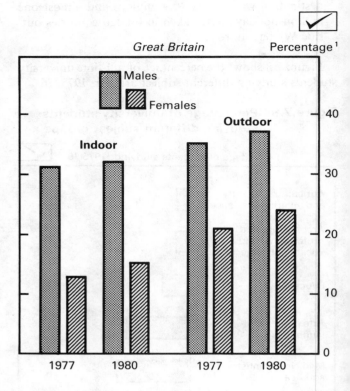

[1] Percentage of the population aged 16 or over engaging in sporting activities in the 4 weeks before interview (annual averages).

Source: General Household Survey, 1977 and 1980

Figure 7.10 Performance of MEPC

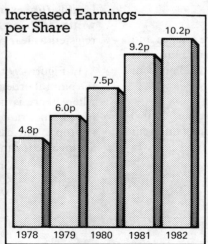

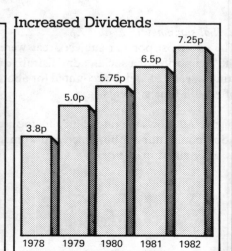

Figure 7.12 Alcohol misuse—admissions[1] to mental illness hospitals and units: by sex and age, 1981

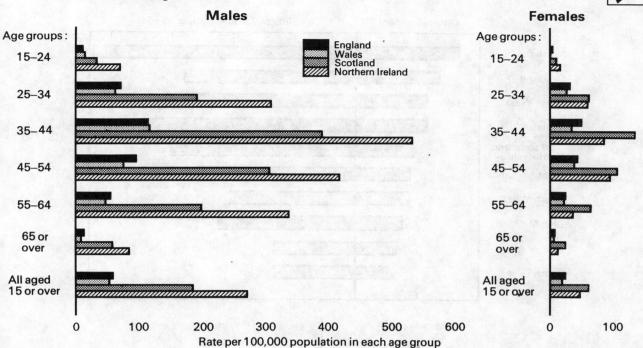

Rate per 100,000 population in each age group

[1] All admissions with a primary diagnosis of alcoholic psychosis, alcohol dependence syndrome, or non-dependent abuse of alcohol.

Source: Department of Health and Social Security, Scottish Health Service, Common Services Agency, Welsh Office, Department of Health and Social Services, Northern Ireland

be drawn narrowly separated as in Figure 7.11 or touching each other as in Figure 7.12.

It is usually best to restrict grouped bar charts to two or three bars per group. With four or more, visual impact is lost unless one set of bars shows a markedly different pattern from the others.

Pairs of related measurements can also be displayed using overlapping bars as in Figure 7.13. This technique works best if the front bar of each pair is consistently smaller than the other.

Correlation between two sets of measurements
Statisticians frequently explore the relationship between two sets of measurements by plotting one measurement against the other in a scatter diagram (see section 7.4). However, non-specialists may not be used to interpreting scatter diagrams and for them a systematic association between measurements may be better illustrated using back-to-back charts.

To do this, the data are first arranged in ascending or descending order of one measurement. This measurement is represented as a series of increasing or decreasing bars.

Figure 7.13 Adult population possessing higher education qualifications: by age and sex, 1981–82

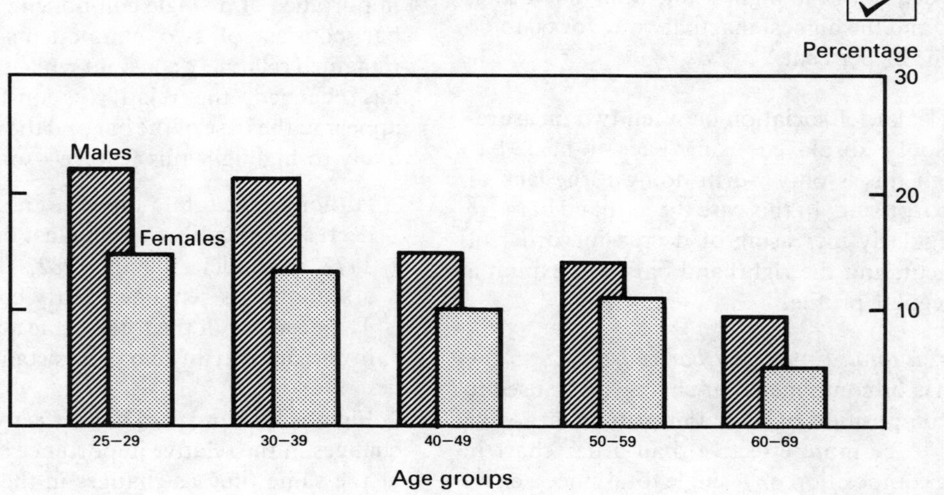

Percentage

Males

Females

Age groups

Figure 7.14 Percentage unemployed and inactivity rates for males aged 60-64: by region

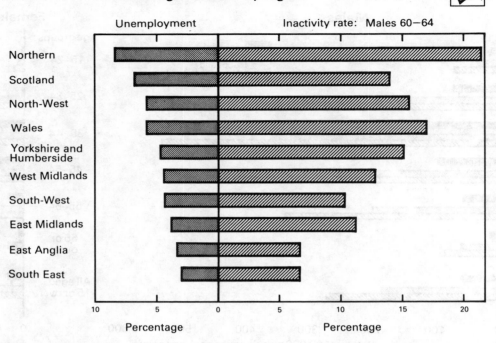

The second measurement is then displayed as bars, each back-to-back with the corresponding first measurement. For example, in Figure 7.14 the data have been arranged in descending order of the original male unemployment rate (Northern at the top, with an unemployment rate of 8.4 per cent; the South East, with 2.9 per cent of men unemployed, at the bottom). The inactivity rate for men aged 60–64 in each region is represented by the righthand bars. The 60–64 year old inactivity rates tend to follow the same pattern as unemployment rates, but with some exceptions. These are highlighted in the verbal summary.

Verbal summary to Figure 7.14
The chart illustrates that in general, areas with high unemployment also recorded high inactivity rates for 60 to 64 year old men. However Scotland, with a higher unemployment rate than the North West, Wales, or Yorkshire and Humberside, recorded a slightly lower inactivity rate than any of these regions. The Northern region recorded both the highest unemployment rate, 8.5 per cent, and the highest inactivity rate for 60 to 64 year old men, 21 per cent.

A complete lack of association between two measurements can also be displayed using back-to-back bar charts, although this is only worth doing if the lack of association is surprising. In this case the lefthand bars are arranged in regularly increasing or decreasing order of one measurement, and the righthand bars will exhibit a completely irregular profile.

Composition of a total: component bar charts
Either pie charts or component bar charts can be used to illustrate the composition of a total measurement. A pie chart is likely to be more effective than a bar chart in illustrating the composition on a single total since a circle gives a stronger impression of a complete entity than does

a bar, but either can be used effectively to illustrate *changes* in the composition of a total. When the compositions of more than three totals are to be compared, component bar charts are likely to work better than pie charts.

There are two ways of drawing component bar charts. The first consists of drawing a series of bars whose height (if drawn vertically) or length (if drawn horizontally) is proportional to the total being represented. Each bar is then sub-divided into components whose sizes are proportional to the components. The second way is to use a *percentage component bar chart* where all bars are the same length, regardless of the total represented, and the chart is used to emphasise changes in the composition of the total.

The first approach should be used with great care. It is best used only to illustrate a marked change in the relative importance of a single component. For example, if each bar consists of two components, one of which has remained relatively constant while the second component has changed, the relatively constant component will appear at the base of the bar and the verbal commentary is likely to highlight this feature—something like:

'although the total number of assistants employed has decreased steadily over the last five years, from 450 in 1977 to under 300 in 1982, the number of male assistants has remained fairly constant at about 180. The decrease in the total is due almost entirely to a fall in the number of female assistants employed.'

Figures 7.15 and 7.16 both effectively illustrate marked changes in the relative importance of a single component at the same time as changes in the composite measurement.

Figure 7.15 Households with current television licences, 1951–1978

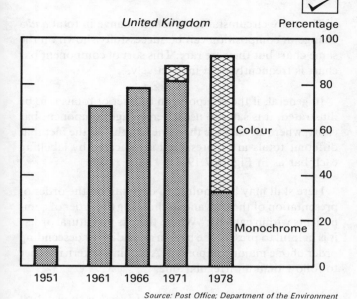

United Kingdom

Source: Post Office; Department of the Environment

In Figure 7.15 we see immediately that the percentage of households with current television licences climbed steadily from approximately 70 per cent in 1961 to over 90 per cent in 1978. The dramatic takeover of colour television between 1971 and 1978 is admirably illustrated by the shading in the last two columns.

In Figure 7.16 there are two clear visual messages: the total stock of dwellings increased steadily over the period 1951 to 1982; and components of the stock have changed

Figure 7.16 Stock of dwellings: by tenure

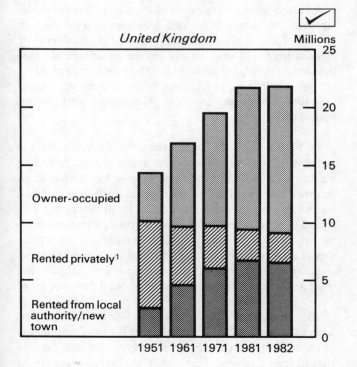

United Kingdom

1 Includes housing associations and dwellings rented with farm or business premises, and those occupied by virtue of employment.
Source: Housing and Construction Statistics, *Department of the Environment*

dramatically. Numbers of privately rented dwellings have fallen while both owner occupation and renting from local authorities have increased substantially.

Unfortunately, component bar charts frequently contain too much information to be effective by including both changes in a total measurement and changes in each of the constituent parts. In general the end result illustrates nothing very clearly. Glance at Figure 7.17 and see what you make of changes in union membership for unions with between 10,000 and 100,000 members from 1961 to 1980.

Figure 7.17 Trade union membership: by size of union

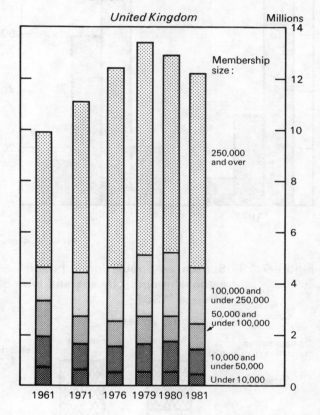

United Kingdom

Source: Department of Employment

In cases like this the recommended approach is to identify the main message in the data and then choose a chart to illustrate it. If the change in constituent parts is of greatest interest, a grouped bar chart can be used as, for example, in Figure 7.18 (overleaf).

From this chart it is difficult to judge what happened to the total number supervised by the probation service from 1971 to 1982 but the slight decline in the number sentenced to criminal supervision, the slight increase in the domestic supervision and after-care categories, and the number in the new category, community service, in 1982 can all be seen easily.

If both changes in total and changes in composition must be highlighted, two simple charts should be considered: say simple column bar charts or a line graph (discussed later) to illustrate the change in the total and two (or possibly three) pie charts demonstrating the

Figure 7.18 Persons supervised by the probation service: by type of supervision, 1971 and 1982

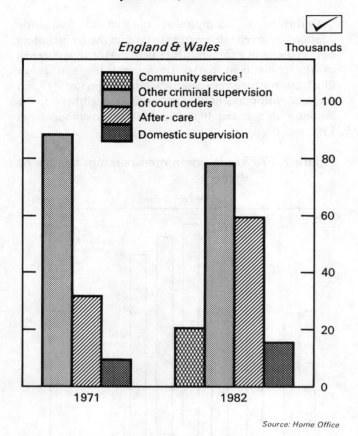

England & Wales   Thousands

- Community service[1]
- Other criminal supervision of court orders
- After-care
- Domestic supervision

1971   1982

Source: Home Office

Figure 7.19 Sentenced population in prison establishments[1]: by sex and type of offence, 1982[2]

England & Wales

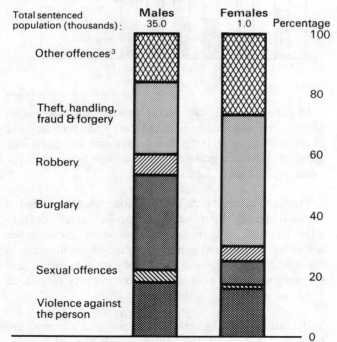

Total sentenced population (thousands):	Males 35.0	Females 1.0	Percentage

Other offences[3]

Theft, handling, fraud & forgery

Robbery

Burglary

Sexual offences

Violence against the person

100
80
60
40
20
0

[1] Includes prisons, borstals, and detention centres.
[2] On 30 June. Includes those imprisoned in default of payment of a fine.
[3] Includes a small proportion of 'Offence not known'.

Source: Prison Statistics, *Home Office*

composition of the total at the beginning and at the end of the period (and, possibly, at a mid-way point).

There are circumstances where a change in total *and* a change in composition can be successfully shown on the same chart but they are rare. This sort of component bar chart is frequently used ineffectively.

In general, if the composition of a set of totals is to be illustrated, it is safer to use a percentage component bar chart, where all bars are the same length and the fact that different totals are represented is indicated by labelling each bar as in Figure 7.19.

Here skill may be required in deciding on the order of presentation of the bars and in choosing the order of components within each bar. Where there is no natural order, it is helpful to present the bars in ascending or descending order of one major component. A regular pattern is easier to follow than an irregular one.

In Figure 7.20 a chart from *Social Trends* is shown, first in its original form, and secondly with the bars rearranged as far as possible in descending order of the 'sole use of all' component. (The exception to this ordering is the 'other' category which remains at the end. This sort of catch-all category often presents problems: it is likely to fall midway through the list if any systematic arrangement of bars is applied, but because it usually consists of a rag-bag of small categories with no dominant component its natural place seems to be at the end of the data. We have not found a tidy solution to this problem, but imposing a systematic ordering on the remaining bars increases the chance of the sequence 'White, West Indian, Indian etc. then African' being remembered.)

An alternative arrangement of the bars is in order of the 'at least one amenity lacked' component as shown in Figure 7.21. This does not solve the problem of positioning the 'other' bar but has the advantage of highlighting a slightly surprising fact: the proportion of white households lacking at least one basic amenity was higher than the proportion of either African or West Indian households.

In Figure 7.21 the components of each bar have a natural order ('sole use of all; some shared none lacked; at least one lacked') and the only decision left is whether to put the 'sole use of all' component at the bottom of each bar or at the top. This chart could be drawn either way round to equally good effect. Where the components have no natural order they should usually be drawn in descending order of magnitude with the largest or most important component at the bottom if vertical bars are used and at the lefthand side if horizontal bars are used.

Taken together the last two guidelines indicate that where there is neither a natural order for the components of each bar nor for the bars themselves, the components should be arranged first, with the largest (or most important) component at the base of each bar, and the bars should then be arranged in order of this component.

78

Figure 7.20   Basic amenities[1]: by ethnic group of head of household, 1977

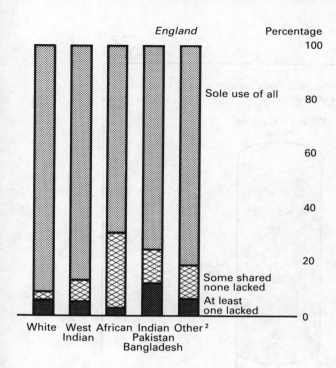

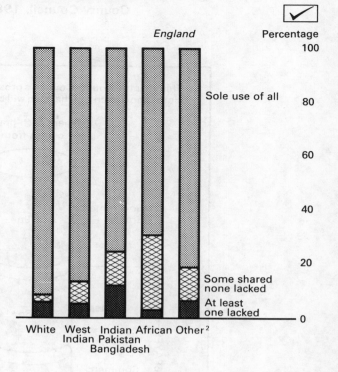

[1] The basic amenities are fixed bath or shower, plumbed hot water supply and wc with entrance inside the building.

[2] Includes Chinese, Other Asian, Arab, Other, and Mixed Origin.

Source: National Dwelling and Housing Survey, *Dept. of the Environment*

Figure 7.21   Basic amenities[1]: by ethnic group of head of household, 1977

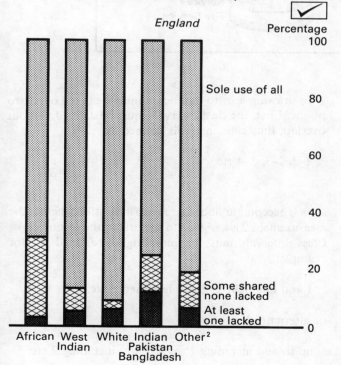

[1] The basic amenities are fixed bath or shower, plumbed hot water supply and wc with entrance inside the building.

[2] Includes Chinese, Other Asian, Arab, Other, and Mixed Origin.

Source: National Dwelling and Housing Survey, *Dept. of the Environment*

Figure 7.22 Composition of income and expenditure of Buckinghamshire County Council, 1982

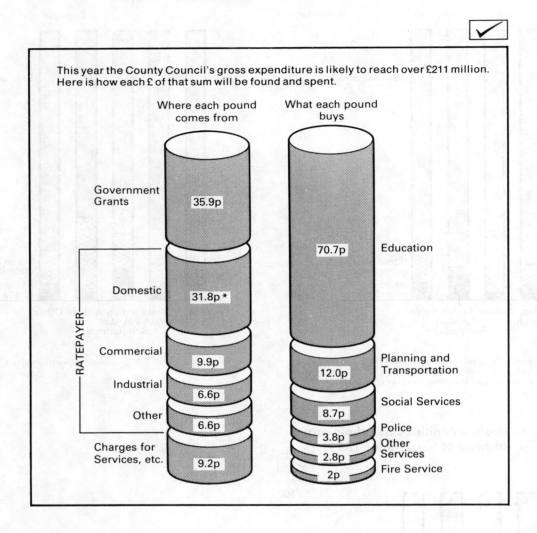

This year the County Council's gross expenditure is likely to reach over £211 million. Here is how each £ of that sum will be found and spent.

Where each pound comes from

What each pound buys

Government Grants — 35.9p

Domestic — 31.8p *

Commercial — 9.9p

Industrial — 6.6p

Other — 6.6p

Charges for Services, etc. — 9.2p

RATEPAYER

Education — 70.7p

Planning and Transportation — 12.0p

Social Services — 8.7p

Police — 3.8p

Other Services — 2.8p

Fire Service — 2p

One common and effective use of percentage component bar charts is in displaying where money came from and how it was spent. **Figure 7.22 (above) shows how Buckinghamshire County Council presented this information to the public in 1982.** Note the device of slightly separating the components in order to make it easier to identify individual components.

Histograms

A histogram is a statistical diagram in the form of a particular kind of column bar chart. Figure 7.23 shows the distribution of radiometric age determinations of samples of rock from around the Arabian Shield.

Technically, a histogram is defined as a series of rectangles each of whose bases represents a range of measurements, called a class interval, and whose area is proportional to the number of measurements falling into the corresponding range. For example, in the histogram in Figure 7.23 the fifth class interval covers the age range 4 to 5×10^8 years and there were 52 samples whose radiometric age was determined as lying in that range.

In drawing a histogram, care must be taken over two points. First, the class intervals must be defined without overlap; thus class intervals defined as:

1–2; 2–3; 3–4 etc.

are *not* acceptable since there is no indication whether the measurement 2 is assigned to the first or the second class. Class intervals must be more tightly defined as, for example:

1 and less than 2; 2 and less than 3 etc.

or alternatively:

up to and including 1; over 1 and including 2 etc.

This point need really only concern the collector of the data. In Figure 7.23 it is not clear what convention has been adopted in defining the class intervals, but the shape of the data is perfectly clear to the reader: the majority of samples had ages determined as lying between 5 and 7×10^8 years, and, on either side of this central peak, the age determinations were more or less symmetrically distributed, with the noticeable exception of a cluster of readings under 1×10^8 years.

Figure 7.23 Distribution of radiometric age determinations of rocks in Saudi Arabia

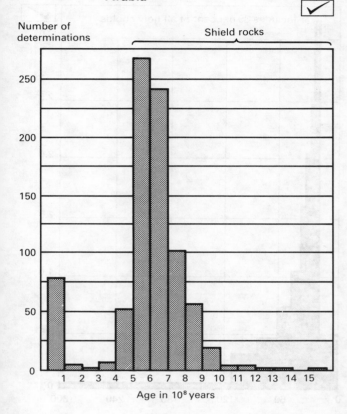

The second point is one of more importance in presenting the data; it concerns the definition of a histogram as a series of rectangles whose *areas* are proportional to the number of observations in each class interval. There is no problem where all the class intervals are of equal width; in such cases, since all the rectangles have equal bases, their heights must be proportional to the number of observations in each class. However, in cases where there are class intervals of different widths, the height of each rectangle must be calculated so that the area represents the number of observations. This can be confusing to readers not familiar with the idea.

In Figure 7.24 below, the bars are of different widths: the first class interval being narrower than most, spanning only four years instead of the more usual five, and there is a wide class interval in the middle spanning 35 to 44 years.

Bars of different widths can distract the reader from the overall pattern, in this case that male unemployment is highest in the 20 to 24 and 60 to 64 age groups and lowest for men between the ages of 35 to 55.

Figure 7.24 Unemployment: by age, April 1982

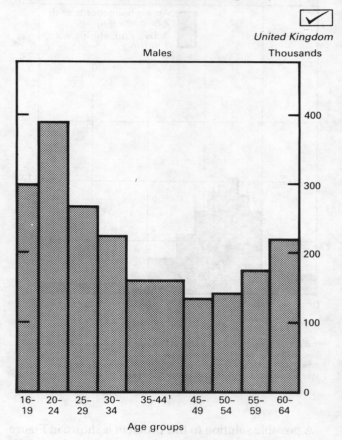

[1] Data are not available separately for age groups 35–39 and 40–44.

Source: Department of Employment

Figure 7.25 Distribution of household income[1]: active households with one and more than one earner, and non-active households, 1977

United Kingdom

Active households[2] with one and more than one earner
Households in each £1 income group as a percentage of all active households

Non-active households[3]
Households in each £1 income group as a percentage of all non-active households

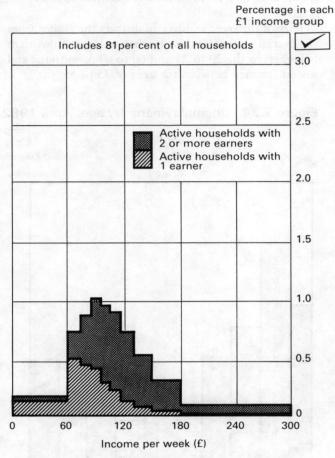

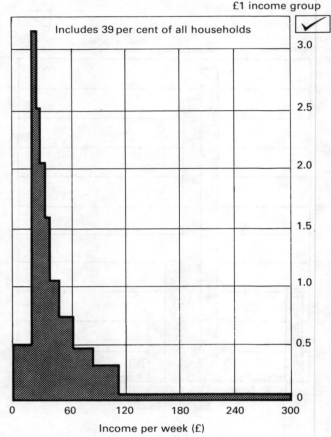

¹ The sum of gross incomes of household members, before income tax and NI contributions, and excluding benefits in kind.
² Where the head is self-employed or in full-time employment.
³ Where the head is employed part-time or is over pension age, retired or unoccupied.

Source: Family Expenditure Survey, *Central Statistics*

A possible solution to this problem is shown in Figure 7.25 (above) where only the outlines of the histogram appear. The reader is no longer distracted by the need to interpret bars of different widths, and can concentrate on the markedly different profiles of the three histograms: in the lefthand chart we see that the most common income range for households with one earner was from £60 to £70 per week, while for households with two or more earners the most common income range was around £100 per week. On the right we see that by far the most common income for non-active households was around £20 per week, and that very few of these households had weekly incomes over £120.

The histogram shown in Figure 7.23 avoided the problem by showing all the bars as being the same width although, strictly, the first bar should have been narrower than the rest since the diagram makes it clear that the first class interval does not start at zero. In this case, statistical exactness has been sacrificed to visual clarity—an approach which may certainly be used, but with caution.

Misuse of bar charts

This lengthy section has illustrated some of the many ways in which bar charts can be used effectively: we have seen bars drawn singly, in pairs, in groups, overlapping, back-to-back, drawn vertically and horizontally, subdivided into components or drawn as piles of money; the lengths of bars have been proportional to absolute magnitudes, to percentages and to the numbers of items falling in specified categories. Unfortunately this versatility can lead to the misuse of bar charts. Because so much information *can* be shown in a bar chart, there is a temptation to cram too much on to a single chart. Figures 7.26 and 7.27 are two examples of bar charts hopelessly overpacked with information. In each case the verbal summary is also included—but in neither case is the reader left with a coherent message built up from the chart and the accompanying text.

Verbal summary to Figure 7.26

Figure 7.26 shows that people of similar ethnic origins tend to live together. There is little current information on where particular groups live, but the pattern of births classified by country of birth of the mother gives some indication. The chart shows the 12 local authorities with the highest proportion of births in 1977 to NCWP born mothers—in each case more than one in four of all births. Nine of these authorities are in Greater London. Together these 12 authorities accounted for 30 per cent of all births to NCWP born mothers in this country. The chart shows how different ethnic groups tend to live together in different areas—although the figures omit births to NCWP mothers who were themselves born in Britain, and include births to women of UK origin who were born in India.

Figure 7.26   Births to mothers born in the New Commonwealth and Pakistan: local authorities with highest percentages, 1977

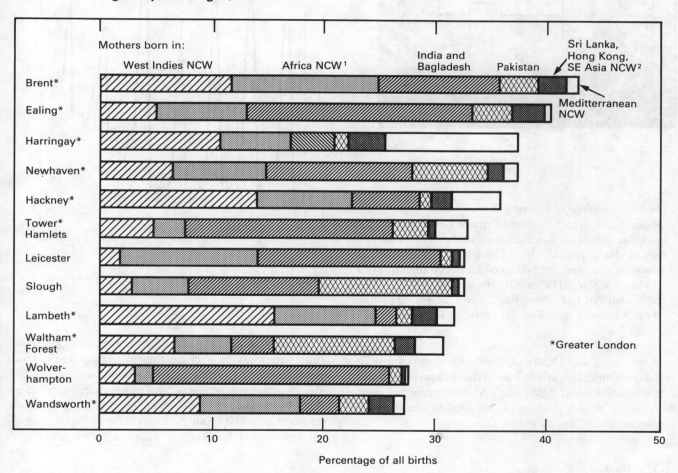

Percentage of all births

[1] Including mothers of Asian descent.
[2] Including other Commonwealth islands in the Indian Ocean, Pacific Ocean, and South Atlantic.

Source: Office of Population Censuses and Surveys

In Figure 7.26 the lengths of bars represent percentages, and have been arranged helpfully in decreasing order of size, *but* each bar has been divided into six portions. The eye registers little more than a confused patchwork. In Figure 7.27, the reader not only has to contend with bars of different heights, each sub-divided into six slices; he also has to interpret bars of different thicknesses. Statistically, the diagram is drawn accurately. Visually it is a mess. (Quickly: without looking at the chart recall *one* interesting fact from the chart.)

Figure 7.27 Male unemployment: by age and duration, April 1979

Great Britain

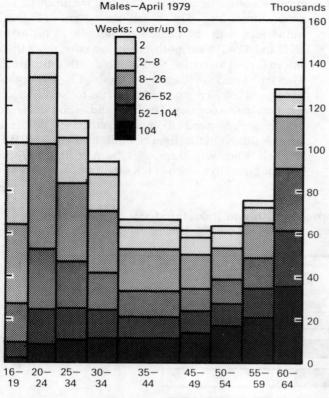

Males—April 1979   Thousands

Weeks: over/up to
2
2—8
8—26
26—52
52—104
104

16— 20— 25— 30— 35— 45— 50— 55— 60—
19 24 34 34 44 49 54 59 64

Source: Department of Employment Gazette

Verbal summary to Figure 7.27

Figure 7.27, which relates to males only, shows that older people are more likely to have been unemployed for a long period. In 1979 almost one half of unemployed men aged 60 to 64 had been unemployed for over a year. The month chosen for the chart has been changed to April so that figures for young people do not include school-leavers later in the year.

When using bar charts, it is extremely important to resist the temptation to 'put it all in the chart in case someone wants to look at it that way'. Always remember your duty to the reader: a chart is provided to illustrate an important point, not as a data store.

7.3 Line graphs

Line graphs can be used effectively to illustrate the movement of a measurement over time. They are quick to draw and easy to understand. Figure 7.28 clearly illustrates both the growth in the number of unemployed school leavers from 1979 to 1983 and the massive annual peak in July, at the end of the school year.

This is an example of a graph representing a process sampled at given times. Strictly speaking, only the points of such graphs are real; the lines joining them are arte-

facts. Those who are worried about such things might prefer to use spike graphs like Figure 7.29, which represents the same data as Figure 7.28.

Spike graphs can sometimes be even more effective than line graphs but are slightly more trouble to draw.

Figure 7.28 Unemployed school leavers[1]

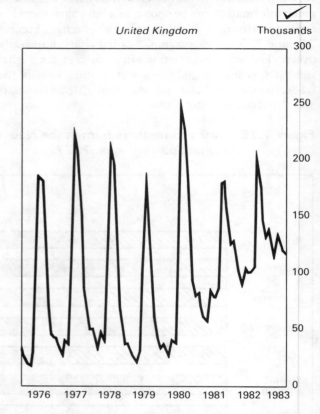

United Kingdom   Thousands

1976 1977 1978 1979 1980 1981 1982 1983

[1] New basis (claimants). School leavers aged under 18.

Source: Department of Employment

In our remaining examples of line graphs, the points represent total or average figures over given intervals (usually years). Once again the lines joining the points are, strictly speaking, artefacts, and purists may prefer to use bar charts instead. They take much longer to draw, however, and often result in a fussier, less effective presentation (imagine, for example, a bar chart version of Figure 7.32).

A point which needs care arises when you wish to break and expand the vertical scale on a line graph. This usually occurs when all the observations lie above a comparatively high value. For example, suppose you wish to show changes in consumers' expenditure on housing between 1973 and 1981 and that the data are recorded based on the 1975 level of expenditure which is taken as 100. All the observed values lie between 95 and 115 so that, if the vertical axis is calibrated from 0 to 115, there is an expanse of blank space at the base of the chart—see Figure 7.30.

Figure 7.29 Unemployed school leavers[1]

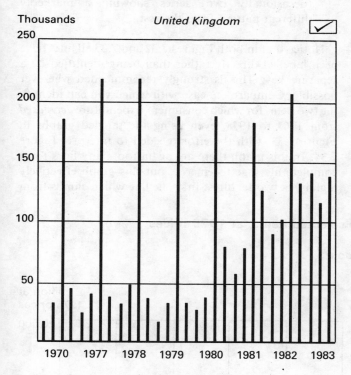

United Kingdom

[1] New basis (claimants). School leavers aged under 18.

Source: Department of Employment

The expanse of blank space below the line may seem unattractive. Actually, it serves the important purpose of enabling the reader to assess at a glance the *proportional* change in consumer spending. If the reader understands (or has been clearly told) that the proportional change is small and the exact pattern of change is of particular interest, the vertical axis may be broken and expanded as in Figure 7.31. This is a device for occasional use in special circumstances and must be employed with care.

Figure 7.30 Consumers' expenditure on housing 1973 to 1981 (1975 = 100)

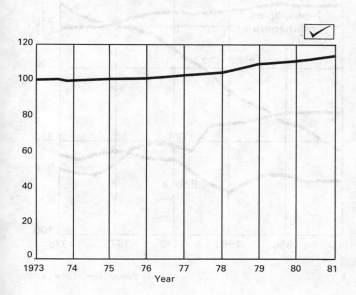

Figure 7.31 Consumers' expenditure on housing 1973 to 1981 (1975 = 100) (Same data as Figure 7.30 presented on different scale)

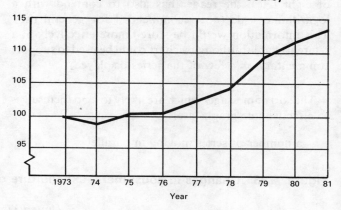

Comparison of growth rates

Line graphs can also be used effectively to compare growth or decline in two or more series of related measurements over the same period, but here great care must be exercised. It is important that the chart should convey a clear visual message and the temptation to put too much on a single chart must be resisted.

Figure 7.32 Changes in relative price indices of domestic fuels[1]

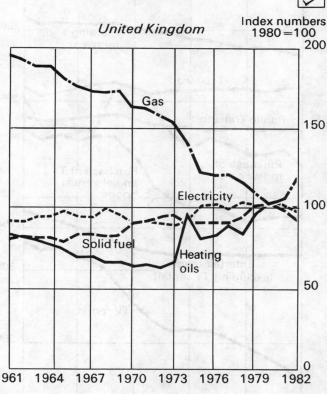

United Kingdom

Index numbers 1980 = 100

[1] Based on the price of each type of fuel, deflated by the fuel and light price index.

Figure 7.32 shows clearly that from 1961 to 1980 the price of gas, relative to other domestic fuels, fell steadily. The eye immediately registers the narrowing gap between the top line and the three lower ones.

By contrast, the two charts shown in Figure 7.33 convey little impression beyond a general upward tangle.

As well as struggling to identify any pattern in two sets of eight lines, the reader has also to contend with a logarithmic scale (not recommended for general use). This information would be stored more effectively in a reference table, although a chart could be used to support commentary on a few of the series displayed.

The sort of messages which are likely to lend themselves to effective line graphs are:

—a number of series moving in parallel

—a small number of series moving in parallel and one (occasionally two) series showing a markedly different pattern from the rest.

Notice that in both Figures 7.32 and 7.33 all lines have been labelled directly rather than being identified by a separate key. This is strongly recommended wherever possible. Compare the ease with which you can identify the two items for which consumers' expenditure *decreased* from 1963 to 1978, even using the tangled graphs in Figure 7.33, with the effort needed to interpret Figure 7.34. In this graph there are again too many lines for a complete message to emerge, but this graph effectively highlights two features: first one line which climbs from

Figure 7.33   Changes in consumers' expenditure on selected items, at 1975 prices

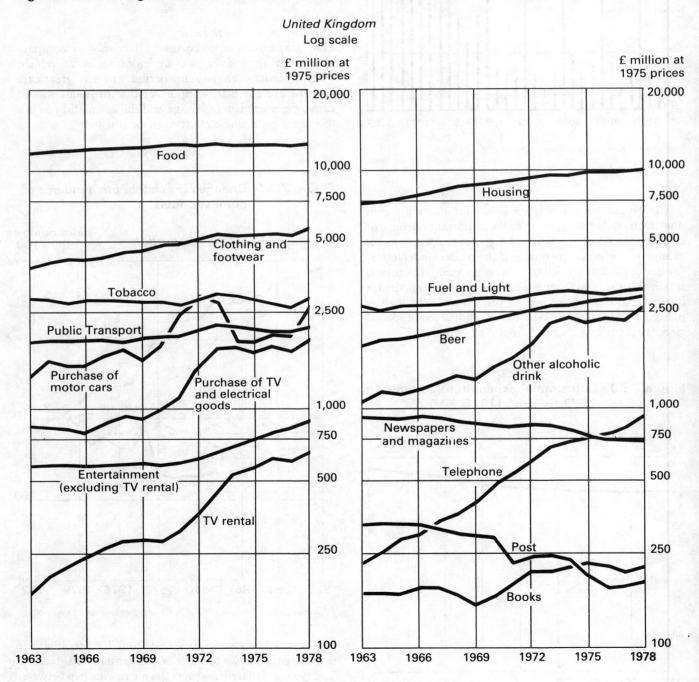

Source: Central Statistical Office

86

Figure 7.34 Divorce rates: international comparisons

Rate per 1,000 existing marriages

England and Wales
Belgium
Denmark
France
Netherlands
Germany (Fed. Rep.)[1]

[1] 1977 and subsequent years are not comparable with earlier years. The First Law Reforming Marriage and Family Legislation came into force on 1 July 1977.

Source: Demographic Statistics 1980 *(SOEC)*

the bottom group to join the top line about 1977, and secondly the downward swoop of a second line between 1976 and 1978.

However the reader has to scan back and forward to the key in order to establish that the first line represents England and Wales and the second represents Germany.

A commonly used variation of line graphs, which cannot be recommended, is a surface or layer chart, such as that shown in Figure 7.35.

In both diagrams, each black line shows the total number of school leavers in any year who had 'these leaving qualifications or better'.

Figure 7.35 School leavers—highest qualification[1]: by sex

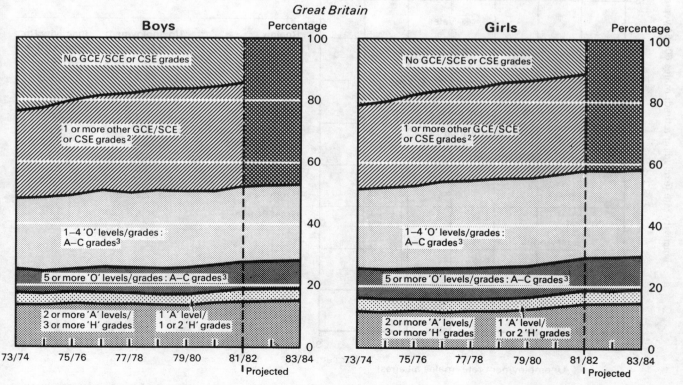

Great Britain

Boys   Percentage

No GCE/SCE or CSE grades

1 or more other GCE/SCE or CSE grades[2]

1–4 'O' levels/grades : A–C grades[3]

5 or more 'O' levels/grades : A–C grades[3]

2 or more 'A' levels/ 3 or more 'H' grades   1 'A' level/ 1 or 2 'H' grades

73/74   75/76   77/78   79/80   81/82   83/84
Projected

Girls   Percentage

No GCE/SCE or CSE grades

1 or more other GCE/SCE or CSE grades[2]

1–4 'O' levels/grades : A–C grades[3]

5 or more 'O' levels/grades : A–C grades[3]

2 or more 'A' levels/ 3 or more 'H' grades   1 'A' level/ 1 or 2 'H' grades

73/74   75/76   77/78   79/80   81/82   83/84
Projected

[1] See Appendix, Part 3: School-leaving qualifications.
[2] Includes 'O' levels /grades- -D and E grades, and CSE grades 2 -5.
[3] Includes CSE grade 1.

Source: Department of Education and Science; Scottish Education Department; Welsh Office.

87

This sort of diagram serves as an easier-to-draw alternative to the component bar chart and can be effective if kept simple. Too often, the presenter is seduced by the ease with which a lot of information can be included in a single diagram and puts in far too much. What happens then is that, as in Figure 7.35, the reader registers little more than a sort of geological cross-section, and looks in vain for the point being illustrated.

Here is the verbal summary which accompanied this chart—but is Figure 7.35 the best way of illustrating this message?

Verbal summary to Figure 7.35
In 1981/82, 55 per cent of school leavers in Great Britain had at least one GCE 'O' level grade A–C or equivalent. The percentage for girls was higher than for boys, 58 compared with 52 per cent. The proportion of

school leavers possessing this qualification in 1973/74 was 50 per cent—48 per cent of boys and 51 per cent of girls.

7.4 Scatter charts

Scatter charts can be used to illustrate a systematic relationship between two measurements (that is, a tendency for an increase in one measurement to be accompanied by a systematic increase or decrease in the second measurement). The chart consists of a horizontal axis calibrated to include the range of one measurement and a vertical axis calibrated over the range of the second measurement. Each pair of measurements is then represented by a single point plotted on the chart. For example, Figure 7.36 shows a scatter chart of inactivity rates among 60 to 64 year old men against unemployment rates. The data from which the chart was constructed are given in the table alongside. Thus the Northern region (with an unemployment rate of 8.4 per cent and an inactivity rate of 21.2 per cent) is represented by a point plotted at 8.4 in the horizontal direction and at 21.2 in the vertical direction.

Figure 7.36   Comparison between regional unemployment* rates and inactivity† rates for males aged 60–64, 1979‡

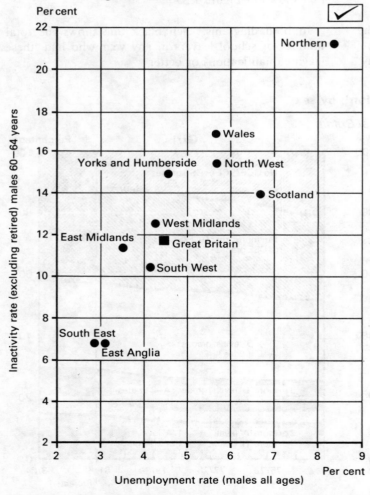

Region	Unemployment Rate (%)	Inactivity Rate (Males 60–64)
Northern	8.4	21.2
Scotland	6.8	14.0
Wales	5.7	17.0
North West	5.7	15.5
Yorks & Humberside	4.6	15.1
West Midlands	4.3	12.8
South West	4.2	10.4
East Midlands	3.6	11.4
East Anglia	3.2	6.8
South East	2.8	6.8
Great Britain	4.5	11.5

* Regional unemployment rates are defined as the percentage of the male labour force aged 16 and over without work and seeking work.   *Source Employment Gazette* 1981
† Inactivity rates for males aged 60–64 are the percentage of the male labour force of those ages reported as inactive but not retired.
‡ Data on unemployment and inactivity rates are derived for the standard regions from the 1979 EC Labour Force Survey.

These data have already been illustrated in this chapter. They were displayed in a back-to-back bar chart in Figure 7.14.

Scatter charts are likely to be less familiar to a general audience than pie charts, bar charts or line graphs. This means that the reader may need some additional help in interpreting the conventions used in drawing the chart and the verbal summary should take account of this. Something like:

'The points plotted in Figure 7.36 represent unemployment rates and inactivity rates for 60 to 64 year old men in ten different regions and in Great Britain as a whole. The chart shows the general tendency for areas with high unemployment to also record high inactivity rates for 60–64 year old men. The Northern region recorded both the highest unemployment rate, 8.4 per cent, and the highest inactivity rate, 21 per cent, for 60–64 year old men. The South East region recorded the lowest rates. Scotland's inactivity rate was low, compared with its unemployment rate, while in Wales the inactivity rate was high.'

Where there are a large number of observations and the strength of relationship between measurements is of more interest than the position of individual points, it is not appropriate to label each point.

Figure 7.36 illustrated two measurements which were *positively* correlated with each other, that is, high values of one measurement tended to be associated with high values of the other. The resultant pattern was a scatter of plots rising upwards to the right of the chart. Where two measurements are *negatively* correlated, that is, where high values of one measurement tend to be associated with *low* values of the other, the pattern of points will slope downwards from left to right.

7.5 Isotypes

Finally, a method of illustrating changes and comparisons which has considerable popular appeal involves the use of isotypes, or pictographs. Here the quantities to be compared are represented by rows or columns of appropriate symbols, for example, simplified silhouettes of men, or ships, or coffee beans. Figure 7.37 shows changes in farm populations in the United States from 1940 to 1973.

The isotype method was developed by Otto Neurath and a team of designers working first in Vienna and later in London over the period 1920 to 1945. ('Isotype' stands for International System of Typographic Picture Education, but the name also implies always using the same symbol.)

Neurath believed strongly in the power of pictures as communication tools and used the isotype system to illustrate economic and social changes in travelling exhibitions and in books like *Modern Man in the Making*.[1] His view was that 'to remember simplified pictures is better than to forget accurate figures'.

Figure 7.37 is essentially a horizontal bar chart, and was taken from a set of three similar bar charts reproduced as Figure 7.38 which illustrates changes in farming in the United States from 1940 to 1973.

---

[1] NEURATH, O. *Modern man in the making*. Knopf: New York, 1939.

Figure 7.37   Changes in farm populations: 1940 to 1973

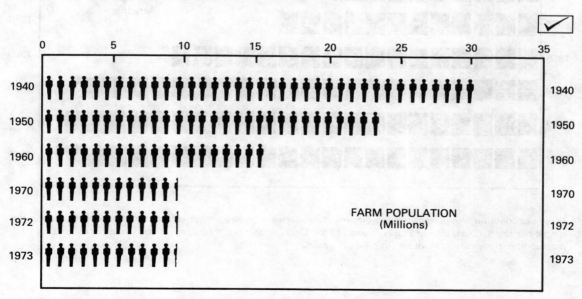

FARM POPULATION
(Millions)

Figure 7.38   Changes in farming: 1940 to 1973

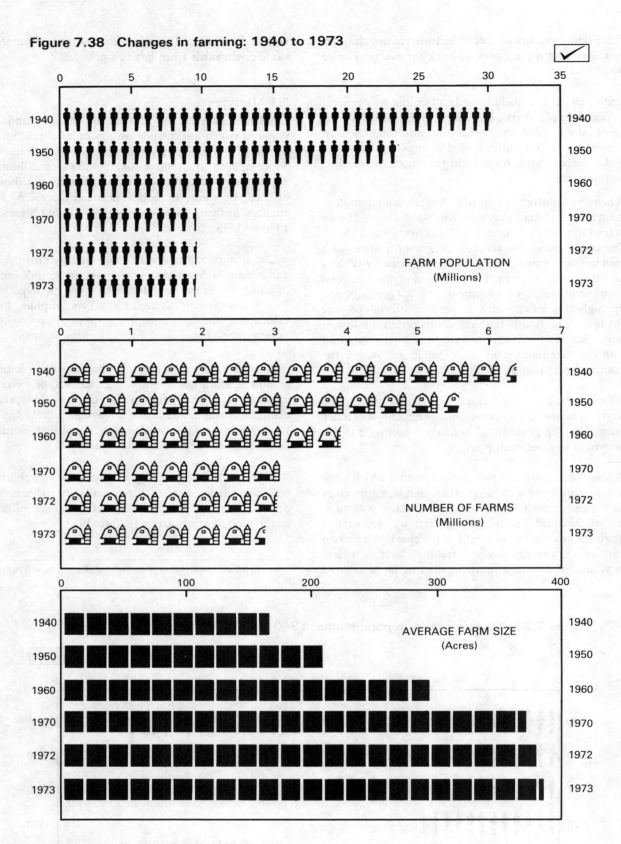

Three pictorial unit charts representing different time series. The basic graphic form for all of these charts is the simple bar chart. *(From United States Bureau of the Census,* Statistical Abstract of the United States: 1974, *Washington: Government Printing Office, 1974, p 592.)*

Figure 7.39 School retention rates

FOR EVERY 10 PUPILS IN THE 5th GRADE IN FALL 1961

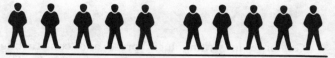

9.6 ENTERED THE 9th GRADE IN FALL 1965

8.6 ENTERED THE 11th GRADE IN FALL 1967

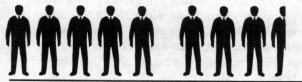

7.6 GRADUATED FROM HIGH SCHOOL IN 1969

4.5 ENTERED COLLEGE IN FALL 1969

2.2 ARE LIKELY TO EARN 4-YEAR DEGREES IN 1973

A pictorial unit chart portraying retention rates, fifth grade through college graduation, United States, 1961 to 1973. Note differences in symbols from one educational level to another. *(From Kenneth A Simon and W Vance Grant,* Digest of Educational Statistics, 1971 Edition, *HEW, Office of Education, Washington: Government Printing Office, 1972, p 8.)*

Isotypes can also be used to construct grouped bar charts or to illustrate a story as in Figure 7.39.

The basic principles of designing an isotype chart and its basic symbols are explained clearly in the Schmids' book on graphic presentation.[1] It is important that:

1. The symbols should be self-explanatory; thus, if the chart is concerned with ships, the symbol should be the outline of a ship.

2. All the symbols on a chart should represent a definite number of units (for example, 1,000 people). Fractions of this number are represented by fractions of the basic symbol (for example, half a symbol for 500 people).

3. The chart should be made as simple and clear as possible. The number of facts presented should be kept to a minimum.

4. Only comparisons should be charted. Isolated facts in themselves cannot be presented effectively by this method.

It is time consuming to produce a clear and professional isotype chart, and a poorly designed or executed one is likely to be inferior to a well executed bar chart. The advantage of using isotypes lies in their ability to 'humanise' the figures. If it is important for your message to appear interesting to the general public and you fear a simple bar chart may not do the job, the use of isotypes should be considered and professional advice sought.

[1] SCHMID, C F and SCHMID, S E *Handbook of Graphic Presentation*, 2nd ed. Wiley, 1979.

Chapter 8: Words

8.1 Don't use words alone

Words alone are very rarely an adequate means of presenting figures. This follows from two well-attested facts:

1. Most numbers are meaningless on their own: their significance derives from comparison with other numbers (for example, the latest monthly unemployment figures assume their full significance only when they are compared with, say, the figures recorded for the previous month or those for the same period last year, or when they are assessed jointly with other economic indicators).

2. It is virtually impossible to include more than two or three numbers in narrative without losing clarity.

The main reason for the difficulty of interpreting numbers embedded in prose is that they are scattered: the reader has to scan the text in order to establish the exact definitions of the numbers before he can attempt to interpret them. Consider, for example, the following sentence:

'Brazil's population has grown at an annual average rate of 2.5 per cent and Malaysia's at 2.6 per cent while real national incomes have increased at annual average rates of 16.5 per cent and 10.5 per cent respectively in the same period.'

There are only four numbers in that sentence. It is easy to pick them out but rather less easy to link them up with their labels. To establish that the 10.5 per cent refers to the annual average rate of increase in real national income in Malaysia takes quite a lot of rapid scanning of the lines of text. By contrast, when data are displayed in a table or a chart, the reader knows exactly where to look for the appropriate information. The table or chart heading gives a general statement of what the data represent. Row and column headings in tables and axis labels in charts give specific definitions of individual numbers. The reader also knows immediately that comparisons can reasonably be made between numbers in the same column or row of a table: physical proximity of numbers in prose is no guarantee that they can be compared directly.

8.2 Vital role of words

However, words have a vital role to play alongside charts and demonstration tables. It is by words that any message is ultimately communicated to its audience. A table or chart without an accompanying verbal commentary is like a silent film: it may be excellently planned and executed, but without words to highlight the important points, indicate general patterns and state the message explicitly, only a few dedicated souls will really understand it.

Guidance in writing lucid prose and constructing memorable reports can be found in other books. For example, the art of writing clear English is well described in Sir Ernest Gowers' *Plain Words*,[1] and *How to Write a Report* by John Fletcher[2] deals helpfully with the problems of writing clear reports. The rules and guidelines discussed there apply to writing about figures as well as to writing about, say, the discovery of North Sea gold. The only difference is that while most people would find the discovery of gold under the North Sea intrinsically interesting they find numbers rather boring and alienating. The writer who has to summarise quantitative information must therefore work hard to make his text as readable as possible.

8.3 Write clearly and concisely

In essence this means that your verbal commentary should be as lucid and as stimulating to the reader as you can make it—while remaining an objective and honest statement of what the numbers say.

Lucidity generally results from an exact understanding of the message you wish to communicate. If you have analysed the data thoroughly and are confident that you know the real patterns and exceptions in the data you will have little difficulty in writing a clear summary. Things to avoid are:

—long sentences, particularly ones with lots of qualifying clauses;

—unfamiliar words: they may be correct; they may be 'stylish', but if there are alternative, everyday words, use them (eg 'alphanumeric data display capabilities' could be rephrased as 'ability to display letters and numbers');

—technical terms, unless your report will be read only by specialists; (eg correlation coefficient, longitudinal survey);

—sloppy or inexact statements (eg use 'represents' rather than 'relates to' and 'results from' rather than 'reflects');

—too many words: the pithier the prose the sooner it will be read and the better it will be remembered.

It can prove difficult to write exactly without becoming long-winded and it is important to ensure that economy of words does not lead to imprecise statements. Be

---

[1] GOWERS, SIR E *The Complete Plain Words*, revised by Sir Bruce Fraser. HMSO, 1973
[2] FLETCHER, J *How to Write a Report*, Institute of Personnel Management, 1983

particularly careful when dealing with changes in percentages and also with proportions of percentages. For example, if you read that 'In 1976, 40 per cent of all applications to industrial tribunals proceeded to hearings while 4 per cent fewer proceeded to hearings in 1977', it is not clear whether the percentages of applications which proceeded to hearings in 1977 was 36 per cent (40 minus 4) or 38.4 per cent (which is 96 per cent of 40 per cent). Here a more appropriate ending to the sentence would be 'while in 1977 the corresponding level was 36 per cent'.

Similarly the sentence 'Between 1978 and 1979 the proportion of applications upheld after proceeding to hearings fell from a quarter to a fifth' leaves the reader uncertain about whether the quarter and the fifth refer to a basis of 'all applications' or to basis of 'those applications which proceeded to hearings'. Since the latter is the correct interpretation, this summary could be rephrased as 'In 1978, 25 per cent of applications which proceeded to hearings were upheld. In 1979, the proportion fell to 20 per cent.'

It is helpful to provide as much structure as possible for the reader, for example, by numbering important points, as in 'Three major trends can be seen over the last fifteen years. They are:

1. . . .

2. . . .

3. . . .'

and by using plenty of sub-headings. Sub-headings form two important functions: first they tell the reader where he has got to in a report ('Ah yes: now we are comparing the UK figures with the rest of the EEC . . .') and secondly they provide memory cues for the reader when he wishes to recall the content of the report ('And then there was the comparison with the rest of the EEC: it said . . .').

A lucid style of writing is essential in all statistical commentary.

8.4 Different sorts of commentaries and different sorts of readers

The best way of making messages interesting to the reader will depend upon who the reader is and why he or she is reading the report. In general there are three sorts of statistical commentary:

1. a verbal summary of a data display (usually a single chart or table) which forms one element of a general report covering non-numerical as well as numerical aspects of a topic;

2. a covering commentary on a table or set of tables published for future reference;

3. a complete report on an essentially statistical investigation.

Each sort of commentary may be read by different sorts of people: the general report may be read by a busy executive who has to use it as a basis for decision-making, or by someone who wants background information on this topic; the covering commentary may be read by a journalist who is going to write a popular article about the figures or by a specialist or researcher who will use the figures in further analysis; the report on a quantitative investigation may be read by the person who commissioned the study, by a statistician or scientist interested in the methods used or by someone with a more general interest in the subject of the report.

Before setting pen to paper, be certain you know whom you are writing for: the busy executive would like all human knowledge summarised on one sheet of A4 (preferably double spaced); the researcher would like a direct and technical explanation of, for example, how the population was stratified and what sort of sampling methods were used; and someone with a general interest in the subject might prefer a journalistic style, with catchy sub-titles and amusing captions.

It is also important to consider the reader's likely attitude to the report and the extent of his or her background knowledge.

If the reader is likely to be:	*Then the report should be:*
interested and knowledgeable about the subject matter, but not a statistician	concise but non-technical
interested but unfamiliar with the background	a little longer: extra tables and commentary may be needed to set your figures in context
apathetic	thought provoking: journalistic tricks may be useful, such as the use of eye-catching sub-titles; relate the report to things the reader knows about, for example by the use of analogies or by referring to well-known events ('This was the year of the Civil Service pay dispute')
hostile or prejudiced	particularly well-structured and thorough: as well as presenting the main conclusions from an analysis of the data, you could include refutations of likely counter-arguments, but a clear structure must be retained
technically expert (a statistician, an economist or other specialist)	concise and exact in its use of technical terms: but if the report may be of interest to a more general audience, it should start with a non-technical summary.

In sections 8.5 to 8.7 below, we deal with the three main types of commentary. There is inevitably some duplication as the considerations which apply to one type of commentary overlap with those which apply to others.

8.5 Verbal summary of a data display

Most of the attributes of a good verbal summary of a table or chart have already been covered in sections 8.3 and 8.4 and in Chapter 4. The following is a list of good practices to follow in writing such a summary; some of the points listed are discussed at greater length in section 8.7 in the context of writing reports.

1. Keep it short: highlight no more than three or four points and never allow the verbal summary to expand into a blow-by-blow account of each entry in the table or chart;

2. Link the summary closely with the table or chart by quoting numbers directly (as in 'From Chart 9.4 we can see that 29 per cent of households had the regular use of one car only in 1961. This proportion increased to 44 per cent in 1971 and then remained more or less constant throughout the 1970's');

3. Avoid emotive or biased descriptions (such as 'shot up by 10 per cent' or 'only rose by a trifling 10 per cent');

4. Unless you are writing specifically for technically expert readers, avoid the use of technical terms (such as 'significant at the 5 per cent level', or 'decreasing marginal rate of return');

5. Do not present changes in definition or other breaks in a series as evidence of trends in the data: the verbal summary should explain what the data reveal of the real situation and should not dwell on the oddities and difficulties of data collection.

Notes or changes in definition or methods of collection should, of course, appear as footnotes and it may sometimes be necessary to refer to such changes in the verbal summary. An introductory sentence along the following lines may then be appropriate. 'From 1977 onwards, Northern Ireland figures are included in the data, leading to an apparent sharp increase between 1976 and 1977 in all categories. However, up until 1976 and from 1977 onwards, the most noticeable features of the data ...'

Examples of clear, concise verbal summaries are given in Chapter 4, pages 44 and 45, and beneath the following table. Table 8.1, is taken from *Social Trends* 13, and the table itself can be criticised: the rows might usefully be rearranged in descending order of the totals (that is, 'Falls', 'Fires', 'Poisoning', 'Suffocation', 'Other') but the verbal summary which accompanies it is admirably lucid.

Verbal summary to Table 8.1

Over half of males and over three-quarters of females who died from accidents in the home and in residential accommodation during 1980 were aged 65 or over (Table 8.1). For young children almost half of total accidental deaths were due to suffocation. About half the total deaths in the 5 to 14 age group were from fires, while poisoning was the most frequent type of accidental death amongst those aged 15 to 44. In the 45 to 64 age group, falls were the most common type of accidental death for both men and women; indeed amongst elderly people, falls accounted for over two-thirds of accidental deaths of men and over three-quarters of accidental deaths of women.

8.6 Commentary on reference tables

Strictly speaking reference tables do not require a commentary: they should be self-explanatory sets of data provided for future use by statistical analysts. However some regularly published reference tables are usually accompanied by a commentary, designed to tell the interested reader about the latest changes in the data. Such a commentary can be provided in either of two ways: each table may be accompanied by its own summary, or a

Table 8.1 Deaths from accidents[1] in the home and in residential accommodation[2]: by sex and age, 1980

Great Britain   Numbers

| | Males | | | | | | Females | | | | | |
	0–4	5–14	15–44	45–64	65 and over	All ages	0–4	5–14	15–44	45–64	65 and over	All ages
Accident type												
Poisoning	5	3	194	98	54	354	6	3	96	102	78	285
Falls	17	4	80	179	883	1,163	7	1	21	144	2,351	2,524
Fires	33	22	63	99	157	374	32	21	48	60	286	447
Suffocation	78	19	99	91	67	354	60	2	21	48	107	238
Other[3]	29	9	59	53	86	236	18	7	25	34	163	247
All accidents	162	57	495	520	1,247	2,481	123	34	211	388	2,985	3,741

[1] Excludes deaths undetermined whether accidentally or purposely inflicted. Data for Scotland include 'late effects'.

[2] Excludes hotels.

[3] Includes accidents caused by explosive material (fireworks, blasting material, explosive gases), hot substances, corrosive liquid, electric shocks, etc.

Source: Office of Population Censuses and Surveys; General Register Office (Scotland)

group of tables may be introduced by a general overview of the trends indicated by them all.

If the first method (a summary for each table) is adopted the commentator's task is comparatively straightforward. All that is needed is a clear, unbiased statement of the latest changes exhibited by the data, and the five points listed in section 8.5 should be observed. Points to bear in mind particularly firmly are:

a. it should be easy for the reader to link up the summary statements with the numbers printed in the tables, and

b. emotive descriptions should be avoided (reject words like 'plummeted'; 'soared'; 'stagnated' in favour of neutral alternatives like 'fell'; 'rose' and 'remained constant': this is a reference table and obvious impartiality on the part of the presenter is essential).

This method is used in the tables published in *British Business*[1] from which Table 8.2 is reproduced from the edition of 23rd July, 1982.

The tables and the associated commentary have a number of strengths and weaknesses. The main weakness of the commentary is that it is difficult to link up the

---

[1] *British Business:* published weekly by the Department of Trade and Industry.

Table 8.2 Extract from *British Business*, 23 July 1982

Production of man-made fibres in March

In the three months ended March 1982 the volume of production of man-made fibres, seasonally adjusted was estimated to have been 4 per cent lower than in the three months ended December 1981. Total production by weight of man-made fibres in March 1982 was 15 per cent lower than in March 1981; output of continuous filament was 25 per cent lower and staple fibre 9 per cent lower than March 1981.
Inquiries: 01–211 7052 or 01–211 4673.

Man-made fibre production

Thousand metric tonnes

	Total	Continuous filament (singles)	Staple fibre
1975	562.5	246.5	316.0
1976	618.4	270.7	347.7
1977	551.8	239.3	312.5
1978	607.2	240.6	366.6
1979	596.3	230.6	365.7
1980	449.8	162.6	287.2
1981	394.6	126.3	268.3
1980			
1st qtr	133.9	51.9	82.0
2nd	125.5	42.4	83.1
3rd	89.0	30.9	58.1
4th	101.4	37.4	64.0
1981			
1st qtr	101.9	38.1	63.8
2nd	102.0	34.2	67.8
3rd	91.6	26.0	65.6
4th	99.1	28.0	71.1
1982			
1st qtr	92.0	28.5	63.5
1980			
January	49.5	19.1	30.4
February	40.5	16.8	23.7
March	44.0	16.0	28.0
April	48.2	16.2	32.0
May	41.8	14.6	27.2
June	35.5	11.6	23.9
July	36.0	11.6	24.4
August	21.2	7.9	13.3
September	31.8	11.4	20.4
October	33.9	12.0	21.9
November	34.0	11.6	22.4
December	33.6	13.8	19.8
1981			
January	32.3	12.0	20.3
February	32.7	12.8	19.9
March	37.0	13.3	23.7
April	36.3	12.4	23.9
May	33.5	11.7	21.8
June	32.3	10.2	22.1
July	30.3	8.9	21.4
August	25.2	7.2	18.0
September	36.1	9.9	26.2
October	33.7	9.9	23.8
November	32.3	9.6	22.7
December	33.2	8.6	24.6
1982			
January	29.6	8.2	21.4
February	30.9	10.3	20.6
March	31.5	10.1	21.4

Source: Production stats: Man-made Fibres Producers Committee. Index of production: Department of Industry.

Index of the volume of production

1975 = 100

1973	129	1976	110	1979	102
1974	111	1977	98	1980	76
1975	100	1978	105	1981	65

	Actual	Seasonally adjusted		Actual	Seasonally adjusted
1980					
1st qtr	92	90	Sep	64	69
2nd	84	80	Oct	68	65
3rd	59	65	Nov	70	68
4th	70	69	Dec	72	74
1981			**1981**		
1st qtr	71	69	Jan	65	69
2nd	68	65	Feb	74	69
3rd	57	63	Mar	73	69
4th	62	61	Apr	74	69
			May	67	62
1982			June	64	65
1st qtr	61	59	July	56	56
			Aug	48	63
1980			Sept	68	72
Jan	98	103	Oct	64	62
Feb	90	85	Nov	63	61
Mar	87	82	Dec	60	60
April	98	93			
May	83	76	**1982**		
June	71	72	Jan	55	58
July	69	70	Feb	66	61
Aug	42	55	Mar	60	57

95

numbers quoted there with the numbers in the tables:

'In the three months ended March 1982 the volume of production of man-made fibres, seasonally adjusted was estimated to have been 4 per cent lower than in the three months ended December 1981.'

This presumably indicates that quarterly figures are being compared, and after a little searching around a badly laid out table, we find quarterly figures: 1st quarter 1982, 59, and 4th quarter 1981, 61; a difference of 2 on 59 which works out as 3.4 per cent. If the numbers are taken from the columns of monthly figures, we find December 1981, 60; March 1981, 57; this gives a difference of 3 on 60, or 5 per cent. It is not clear from where the 4 per cent comes.

The next statement: 'Total production of man-made fibres in March 1982 was 15 per cent lower than in March 1981' is easier to reconcile with the tabulated numbers. Clearly the numbers being compared are 37.0 (March 1981) and 31.5 (March 1982): their difference is 5.5 which is indeed 15 per cent of 37 (to the nearest whole number): but the reader would have found it easier to link the text to the table if this statement had included the tabulated numbers. For example, the commentary might have said 'Total production by weight of man-made fibres in March 1982 was 15 per cent lower than in March 1981 (31,500 metric tonnes as opposed to 37,000 metric tonnes) ...'

By contrast, Table 8.3 (opposite), taken from the same edition of *British Business*, is accompanied by an exemplary verbal summary. The most striking of the latest figures are commented on succinctly and all the numbers mentioned in the summary can be easily identified in the table.

Verbal summary to Table 8.3
Domestic furniture deliveries of manufacturing establishments employing 35 or more people in May are provisionally estimated at £66.1m (at current prices), which gives a seasonally adjusted index of deliveries for the month of 82. The average for the latest three months, March to May is 76, which is 2.5 per cent higher than the figure for the previous three months (December to February) and 11.7 per cent lower than the figure for the same period in 1981.

The index of orders-on-hand, on a seasonally adjusted basis, yields a provisional figure of 39 for May. The average for the three months March to May is 43 which is a fall of 4.6 per cent from the figure for the previous three months, and 16.3 per cent lower than the corresponding period last year.

If a statistical commentary is designed to cover a number of reference tables, as for example the three or four-page commentary which typically accompanies the Labour Market Data section in the middle of the *Employment Gazette*,[1] then a slightly different approach is required. Here the commentator's task is to provide an overview of the latest data in a number of related fields: it is reasonable to assume that the reader will be well informed about previous trends but, apart from that, the commentary will have many of the same characteristics of a report on a statistical investigation. Writing such a report is discussed below.

8.7 Report on a statistical investigation
The topic of report writing as a whole is beyond the scope of this book but some points of specific relevance to reports on statistical topics are listed below:

a. In writing a report of a statistical investigation, start with the main findings. Every journalist knows that many more people read the first paragraph of any article than read the last. The same applies to statistical reports. By starting with the main findings you ensure that, when the reader is interrupted by the telephone or the need to prepare for a meeting in ten minutes' time, at least the most important points have been read. Also, if they were interesting and lucidly written, the reader is more likely to want to read the rest of the report.

b. Wherever possible, start a report with clear, universal conclusions. You may have to resist a temptation to give first place to some particularly dramatic findings which applied only in limited circumstances. For example, in comparing the fuel consumption of petrol and diesel powered cars, the diesel vehicles might use 30 per cent less fuel in stop-start town driving but only 5 per cent less fuel on steady motorway cruising at 50 miles per hour. Here it would be better to start with the weaker statement 'Under all the driving conditions tested the diesel vehicles consumed at least 5 per cent less fuel than the petrol driven vehicles'. You can then continue the report by explaining that, in some circumstances, the savings in fuel were considerably greater than 5 per cent.

If you start with the dramatic savings of 30 per cent in town driving and then weaken the message by saying that, under other conditions the fuel savings were less noteworthy, the reader instantly suspects that, if he or she reads on far enough, the advantageous performance of the diesel powered cars may vanish entirely.

c. The commentator must strive to be objective. This can present problems because the selection of summary statements is inevitably a subjective process. However, if the clearest patterns in the data are the ones selected for comment, then subjective bias will be minimised. By the same token, descriptions should be couched in neutral terms: downward trends should not 'plummet' or 'hurtle'; upward trends should not 'soar' or 'rocket'. Regrettably,

---

[1] *Employment Gazette*: published monthly by the Department of Employment. The middle section is printed on orange paper and contains tables of data on Employment, Unemployment, Vacancies, Industrial Disputes, Earnings and Retail Prices.

Table 8.3 Domestic furniture deliveries and orders, May 1982

	Deliveries £m current values[2]	Index of deliveries (1975 = 100)		Index of orders[1]	
		Not s.a.	s.a.	Not s.a.	s.a.
1975	578.8[3]		100		100
1976	660.1[3]		102		87
1977	751.7[3]		99		75
1978	911.1[3]		108		85
1979	1 053.0[3]		111		91
1980 {	1 005.3[3]				
	992.9		94		59
1981	955.4		85		51
1981					
1st qtr	255.4	93	90	55	53
2nd	228.6	82	86	48	51
3rd	221.6	79	87	53	55
4th	249.8	88	82	49	47
1982					
1st qtr	217.3	75	72	47	45
1981					
Jan	80.6	90	95	60	55
Feb	84.9	92	90	56	54
March	89.8	97	87	49	52
Apr	79.0	85	86	45	47
May	65.2	70	85	50	54
Jun	84.5	90	87	48	54
Jul	71.2	76	92	52	57
Aug	66.0	70	89	55	54
Sep	84.4	90	79	56	53
Oct	92.9	98	85	49	47
Nov	86.2	91	81	49	44
Dec	70.8	75	79	45	44
1982					
Jan	64.0	67	74	48	45
Feb	69.1	71	70	48	46
Mar	84.3	86	74	42	45
Apr	69.6	71	72	39	41 r
May p	66.1	67	82	36	39
% change latest					
3 months on					
prev. 3 months			+2.5		-4.6
On a year earlier			-11.7		-16.3

[1] Orders on hand at end of month or centred average (average 1975 = 100).
[2] Deliveries of establishments employing 35 or more people, except where marked [3].
[3] Deliveries of establishments employing 25 or more people. p Provisional. r Revised.

Inquiries: Department of Industry, Statistics Division 1C,
Room 1918, Millbank Tower, Millbank, London SW1P 4QU
Tel 01-211 4028.

Note

The indices are given on 1975 constant price basis, with 1975 = 100. When price increases are large, the problems associated with revaluation to constant prices become more acute. Thus the relative accuracy with which volume changes have been measured is likely to have been reduced in times of high inflation.

this necessary asceticism can lead to flat and boring prose. The best safeguards against this are a crisp, clear style and a well-structured report.

d. It must be easy for the reader to link the points discussed in the commentary with the relevant reference tables. This can be achieved in two ways: first by quoting directly from the tables, as in 'The rate of inflation, as measured by the 12-monthly change in the Retail Price Index, continues to slow down with a further reduction to 6.8 per cent in October (see Table X.Y). This compares with 7.3 per cent in September, 8.0 per cent in August and 12.0 per cent at the beginning of the year'. (Needless to say, the numbers 6.8, 7.3, 8.0 and 12.0 must all appear in Table X.Y.)

The second way of linking the commentary to the tables is by using sub-headings.

e. The reader will find short sections of commentary, each with a distinct sub-heading, much more digestible than a single length of continuous prose. The sub-headings should indicate which tables are commented on in the following section: for example, a heading 'Industrial stoppages' would include comments on tables giving the numbers of industrial disputes and stoppages of work. So long as they maintain the structure of the report clearly, eye-catching subtitles can be extremely effective. Questions, quotations and puns can all be used with the effect of jolting the reader's attention and thus producing landmarks in the report. The following examples are all from the same edition of *The Economist*, a publication which specialises in presenting lucid quantitative analyses: 'For whom the gong tolls'—above an article on the Rank Organisation; 'Ethics and ethnics' above an article on Bank Bumiputra (an offshore Malaysian bank of doubtful reputation); and, as a heading above a chart showing the sharp collapse of Polly Peck (Holdings), 'Sukey, take it off again'. Used skilfully and conservatively captions can enliven the most prosaic report.

f. Consider using small additional data displays (charts or tables) where necessary to demonstrate important points: for example, line graphs to illustrate trends over time and well planned demonstration tables to bring together figures from different reference tables. This may also be done if the reference tables are to be set against a wider background, for example, by giving comparable rates for other EC countries.

g. Do not give detailed explanations about changes in definitions or incompatibilities within the data as part of the commentary: they belong as footnotes.

h. Do not put forward changes in the method of data collection as 'reasons' for (observed) patterns. It may be desirable to comment on such a change, usually to prevent the data being misinterpreted—but the comments must be carefully worded. Thus 'The increase in 1982 over 1981 is due to a change in the reporting system' should be rephrased along the lines of 'The new reporting system in 1982 caused reported numbers to increase. It is thought that, under the same reporting system, the 1982 numbers would have been about the same as those recorded in 1981.'

i. Avoid technical terms unless you are writing only for a specialist audience. Where a technical term has to be used because of its unique ability to describe an important point, include a non-technical description of its meaning either in the text, if this can be achieved smoothly, or as a footnote, as well as giving a precise definition in the list of definitions at the end of the report. For example, in an article on regional accounts,[1] the following paragraph introduces the non-technical reader to the phrase 'structural component':

'A question which is of perennial interest in considering the economic problems of the regions is the extent to which overall differences in rates of change are due to the effects of industrial structure on the one hand (the "structural component") or to differences between regional performances within each industry on the other (the "growth component"). An adverse structure would be one which had, for instance, an above average share of typically low-growth industries.'

8.8 Example of a good commentary on statistical tables

The final pages of this chapter are taken from a quantitative report on Civil Service Statistics.[2] The paper, written in 1980, was produced as background to consideration of the effect of the Government's commitment to reduce the number of civil servants to 630,000 by 1984.

The tables and graphs are reproduced here in their original form and do not follow all the guidelines suggested in this book (for example columns of Table 1 are too far apart; the graph could have been reduced to half its size and printed vertically in half a page, the calibrations on the axis could have been closer to the axes and the vertical axis could have had small marks at the calibration points). However, the commentary which accompanies them, in paragraphs 3 and 4, is clear and helpful. The graph has been included because, in general, time trends are easier to see when displayed as line graphs than from a table, as recommended in section 8.7 (f).

In reading the brief commentary (paragraphs 3 and 4 from the extract) note the following virtues:

—The reader is immediately referred to the appropriate table and told that the same data have been displayed graphically;

---

[1] 'Regional Accounts': an article in *Economic Trends* no 349, November 1982, HMSO, pp 82–98.

[2] House of Commons Library Research Division. *Civil Service Statistics*, by Robert Twigger, 1980. (Background paper no 85.)

—The summary starts with a clear statement of the broad, overall picture: numbers fell from a war-time peak to 643,000 in the period 1960–61 and then increased until 1976. This general pattern is immediately seen from the graph—but, for closer scrutiny, the numbers can readily be found in Table 1. More detailed comments (about differences between industrial and non-industrial staff) follow next;

—The style of the commentary is crisp and factual: no emotive adjectives or verbs are used;

—The report is well structured, with sub-headings used to indicate a change of topic and consideration of a different table;

—The estimated effect of changes in definition on the true position is stated clearly (if slightly ungrammatically) in paragraph 4 ('The true fall is probably closer to 34,100 due to changes in which staff are included . . .') and the reader is referred to the relevant footnotes for further explanations.

—No technical terms are used and the patterns in the data are related to facts with which the reader is likely to be familiar: '. . . as a result of the temporary ban on recruitment announced on 22 May'.

Extract from Civil Service Statistics

3. Civil Service Manpower since the War

Table 1 shows the number of civil servants at 1 April each year 1939 and 1944–1980, this data is also graphically represented (page 101). The number of civil servants fell from the war-time peak reaching 643,000 in the period 1960–61. However after 1961 the trend was reversed and the total rose during the sixties reaching a new peak of 748,000 in 1976 since which time the total has been falling again. The increase in the number of civil servants during the period 1961–1976 was entirely due to growth in the non-industrial Civil Service. During this period the number of industrial servants fell steadily and the industrial Civil Service shrank from 40% of staff in 1961 to 24% in 1976. In fact, apart from a short period in the early fifties the number of civil servants in industrial roles had declined steadily since the end of the War. There have been numerous changes in the groups which are counted as civil servants; major changes are detailed in the footnotes to Table 1.

4. The current position

Table 2 shows the number of civil servants analysed by ministerial responsibility each quarter from April 1979 to October 1980. Over this period total manpower has fallen by 35,200. 22,600 non-industrial posts have disappeared, 4.0% of total staff in April 1979 and 12,600 (7.6%) industrial posts. The true fall is probably closer to 34,100 due to changes in which staff are included (see footnotes (c) and (g)). The largest cuts were in environment, defence and the Inland Revenue whilst employment in the Home Office increased. Over half of the fall in the total during this period occurred in the six month period up to October 1979 when 20,000 posts were cut mainly as the result of the temporary ban on recruitment announced on 22 May.

The Government's planned reductions are on a net basis, so that while the total number of civil servants is due to fall there will be certain departments which will have an increase in staff (see Fourth Report of the Treasury and Civil Service Select Committee 'Civil Service Manpower Reductions' HC 712–1 of 1979–80, para.7). The plans allow for an increase of 11,500 posts in the DHSS and Home Office during 1980–81 to cope with extra payment of benefits and additional manpower for law and order.

Table 1   Numbers of industrial and non-industrial civil servants 1939–1980 (a)
(United Kingdom at April each year)

Year	Non-industrial (000's)	Industrial (000's)	Total (e) (000's)
1939	163	184	347
1944	505	658	1,164
1945	499	615	1,114
1946	452	366	819
1947	457	326	784
1948	445	317	761
1949	458	326	784
1950 (b)	433	313	746
1951	425	316	740
1952	429	333	762
1953	414	341	756
1954 (c)	405	347	751
1955	386	334	719
1956	384	328	711
1957	381	314	696
1958	375	289	664
1959	375	271	647
1960	380	263	643
1961	387	256	643
1962	394	253	647
1963	410	252	662
1964	414	244	658
1965	420	235	655
1966 (d)	430	232	662
1967	451	229	680
1968	471	222	693
1969	470	214	684
1970 (a)	493	208	701
1971	498	202	700
1972	496	194	690
1973	511	189	700
1974 (f)	512	180	692
1975 (f)	524	177	701
1976	569	179	748
1977	571	174	746
1978	567	168	736
1979	566	167	732
1980	548	157	705

Notes

(a) Full-time-equivalents excluding Post Office staff throughout (except staff of the Department of National Savings and the former Ministry of Posts and Telecommunications which are included from 1970). Casual staff and period appointments are also omitted.

(b) 10,000 staff of approved societies and local authorities were transferred to the Ministry of National Insurance in 1949–50.

(c) In 1954, 7,000 non-industrial and 10,700 industrial staff were transferred from the Civil Service to the UK Atomic Energy Authority.

(d) Casual, non-industrial staff are included before 1969.

(e) Totals may not be the sum of the figures shown due to rounding of the data.

(f) Figures for 1974 and 1975 exclude employees of the Manpower Services Agency which took over certain functions from the Department of Employment in January 1974 but whose staff were not counted as civil servants until January 1976.

Sources

Civil Service Statistics 1971 page 14
Monthly Digest of Statistics various issues table 3.4
Civil Service Department.

Civil Service manpower 1945–1980

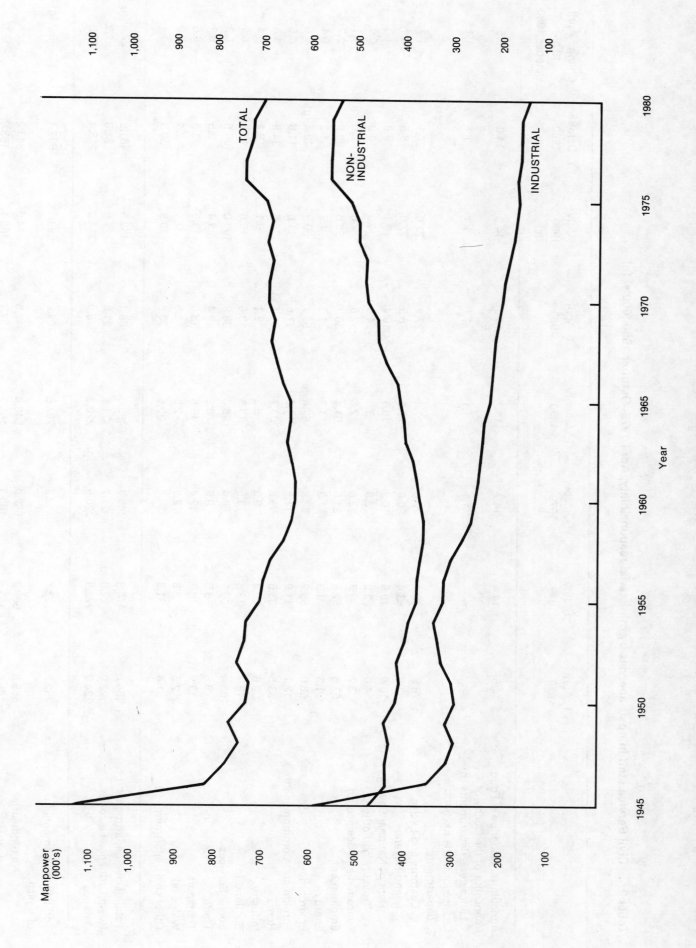

Manpower
(000's)

TOTAL

NON-
INDUSTRIAL

INDUSTRIAL

Year

Table 2 Civil Service staff in post: analysis by ministerial responsibility, April 1979–October 1980 (000's) (a)

	1 April 1979	1 July 1979	1 October 1979	1 January 1980	1 April 1980	1 July 1980 (c)	1 October 1980	Change April 1979–October 1980
Agriculture, Fisheries & Food	14.5	14.4	14.2	14.2	14.3	14.2	14.0	– 0.5
Chancellor of the Duchy of Lancaster's departments (b)	–	–	1.1	1.1	1.1	1.1	1.1	+ 1.1
Chancellor of the Exchequer's departments:								
Customs and Excise	28.8	28.4	27.8	27.4	27.2	27.0	27.1	– 1.7
Inland Revenue	84.6	83.4	80.6	79.0	78.3	77.0	76.5	– 8.1
Dept of National Savings	10.8	10.5	10.2	10.3	10.4	10.1	9.8	– 1.0
Treasury and others	4.0	3.9	3.9	3.9	4.0	4.0	4.0	0.0
Education and Science (b)	3.7	3.7	2.6	2.6	2.6	2.6	2.6	– 1.1
Employment	53.6	52.6	51.6	51.4	50.7	50.6	50.9	– 2.7
Energy	1.3	1.3	1.3	1.3	1.3	1.2	1.2	– 0.1
Environment	56.0	55.2	53.8	52.6	51.7	50.9	49.4 (g)	– 6.6
Foreign and Commonwealth	12.1	11.9	11.8	11.6	11.6	11.6	11.6	– 0.5
Home	33.5	33.5	33.5	33.7	34.1	34.6	34.9	+ 1.4
Industry	9.5	9.4	9.3	9.2	9.1	9.1	9.1	– 0.4
Scotland (d)	13.7	13.6	13.5	13.5	13.6	13.6	13.6	– 0.1
Social Services	100.9	100.1	98.4	98.0	98.9	100.0	100.6	– 0.3
Trade	9.6	9.5	9.6	9.5	9.4	9.4	9.5	– 0.1
Transport	13.9	13.7	13.4	13.5	13.5	13.5	13.3	– 0.6
Wales (e)	2.6	2.6	2.5	2.5	2.5	2.4	2.4	– 0.2
Other civil departments (f)	31.4	31.3	30.6	30.6	30.9	30.6	30.3	– 1.1
Total of civil departments (h)	484.6	478.9	469.7	465.9	465.1	463.5	461.8	–22.8 (g)
Royal Ordnance Factories	23.0	22.0	21.7	21.5	21.8	21.6	21.8	– 1.2
Defence	224.7	222.9	220.9	220.2	218.0	214.8	213.4	–11.3
Total all departments (h)	732.3	723.8	712.3	707.6	704.9	699.9	697.1	–35.2 (g)
of which:								
Non-industrial	565.8	560.2	552.0	548.4	547.5	544.9	543.2	–22.6
Industrial	166.5	163.5	160.3	159.2	157.4	155.1	153.9	–12.6

See footnotes and answers opposite

Notes

(a) Part-time staff counted as half units.

(b) The Office of Arts and Libraries under the Chancellor of the Duchy of Lancaster formally came into operation in September 1979 and took over responsibility for the Victoria and Albert Museum and the Science Museum from the Department of Education and Science.

(c) About 200 period appointed staff previously excluded from the count are included from July 1980.

(d) Departments of the Secretary of State for Scotland and the Lord Advocate.

(e) Welsh Office.

(f) Including the Northern Ireland Office which is part of the UK Civil Service but excluding the Northern Ireland Civil Service.

(g) 1,276 Property Services Agency staff previously included in the Manpower Count are excluded from October 1980.

(h) Totals may not be the sum of the figures shown due to rounding.

Sources

Monthly Digest of Statistics 1980 table 3.4 and earlier editions.
CSD *Summary Quarterly Staff Return* 1 October 1980.

Appendix A: Tables and charts as visual aids

Introduction

The writer of a report is frequently asked to present his or her findings orally. If the report includes statistical analysis, this presentation is likely to require slides, wall charts or viewgraphs (overhead projector slides) of the key charts and tables.

This appendix does not include a technical description of how to prepare slides (for that the interested reader is referred elsewhere)[1]. It does, however, offer guidance for those with little experience in the use of visual aids who cannot always call on the services of a profesionally trained graphics officer. Most of the guidelines have been developed from personal experience over a number of years at the Civil Service College and elsewhere, and from both sides of the lecturer's table.

Viewgraphs

The easiest visual aid to prepare and use is the viewgraph or overhead projector slide (OHP slide). The table or chart to be displayed is drawn on an A4 sheet of acetate using either permanent or water-soluble pens, or printed directly onto the acetate using a heat copying machine (but see section 7, below, for possible dangers). For projection, the slide is placed right way up on a projector enabling the speaker to read text from the slide without turning away from the audience. The speaker can point to features of particular interest on the slide using a thin pen or pointer and can add lines or further numbers to the slide during the lecture. There is no need to lower the lights in a lecture room in order to use an overhead projector and, if no screen is available, the image can be projected quite satisfactorily onto a blank wall.

The overhead projector is therefore an extremely valuable lecturing aid. Modern overhead projectors are compact and portable and are automatically provided in many lecture rooms. (But always check in advance.)

Production of viewgraphs

If you plan your presentation well in advance and can call on the services of a graphics officer, your talk can be illustrated with professionally produced viewgraphs. This is obviously ideal. Particularly if you are likely to give the same presentation several times, it is well worthwhile getting a set of expertly executed illustrations.

It is however perfectly possible to produce effective, if amateur, viewgraphs yourself if some simple points are

carefully observed. Even when the final slides will be drawn by an expert, it is you, the speaker, who must plan the content and general design of each slide. A good graphics designer may offer advice about detailed execution, but only the speaker knows exactly what message each slide must contain. The first six points in the following list are therefore applicable both when you intend to draw the slide yourself and when you will leave the production to a professional.

1. *The slide must be readable at the back of the lecture room.* This means:

   —do not put too much on to a single slide: about six lines of well-spaced text or numbers will usually be legible

   —use bold letters and numbers

   —use clear colours for numbers and text (never yellow or orange).

2. *It must be intelligible.* This means that it must not contain too much information. One simple message per slide is the general rule, which can, however, be modified by using overlays and reveals to build up a more complex message. (See 3, 4 and 5 below.)

3. Overlays can be used effectively to develop an argument. These consist of one or more extra sheets of acetate hinged to the original viewgraph so that they line up exactly with the text or chart on the original slide. They can be used to:

   —superimpose new points, lines or bars on an existing chart

   —produce an extra row or column of figures to supplement a table

   —put summary captions on a slide

   —build up a story caption by caption.

4. Blank space can be left on the slide to be filled in during the talk; for example, the three latest points might be added to a time-trend graph, or a row of average figures might be added to a table of numbers. This approach can be particularly effective if the information added to the chart shows a striking pattern and is likely to produce a livelier, more participative talk than the use of pre-prepared overlays. (Naturally you must have prepared this in advance and be sure that you know exactly what to add and where to write it on the viewgraph: you must also have checked that suitable, working, OHP pens are available during the lecture.) If the slide has been executed in permanent ink and a water-soluble pen is used to fill in extra numbers or lines, additions can be erased later. Alternatively

---

[1] For example, American Society of Mechanical Engineers. *Illustrations for Publication and Projection.* American National Standards Institute. Y15.(1M–1979).

additions can be written on a blank overlay, which prevents any portion of the original slide being accidently erased.

5. Alternatively reveals can be used to build up a story caption by caption. Here all the captions are printed on the same slide which is initially masked so that only a small section is visible to the audience. As a new point is developed in the talk, the mask is progressively removed to reveal the next caption.

6. Use colour constructively. Coloured viewgraphs are always more attractive than black and white ones, but this should not lead to meaningless changes of colour merely for the sake of variety. Different colours can be used in a number of constructive ways. For example, black can be used for tables and diagrams which are reproduced in handouts which accompany the talk, while additional slides or portions of slides are drawn in another colour; in tables, red figures can be used for female numbers and blue figures for male numbers; if summary captions follow a table, the colours of captions can be matched to the colours used for the columns (or rows) of the table; tables can be drawn in one colour, captions in another.

7. *Do not produce viewgraphs by xeroxing standard typing or printing.* Many inexperienced lecturers discover with joy that viewgraphs can be produced by a heat copier from photocopies or from printed tables and charts. This can seem like the ideal way of illustrating a talk with rapidly produced, professional looking slides. It is actually a recipe for producing over-crowded, illegible slides and a restive audience. Standard typeface is too thin and cramped for use on overhead projector slides. If you do not have sufficient time to get slides produced using a special enlarged typeface or on a headliner machine, a slide drawn clearly and lettered by hand, using proper overhead projector pens, will be infinitely better than a slide reproduced from a typed page.

8. If you intend to draw your own slides it is important to lay them out tidily and to make your letters and numbers as regular as possible. The layout should be carefully planned on squared paper (ordinary graph paper can be used) to ensure that columns are vertical and evenly spaced and that rows are regular and horizontal. This paper should then be placed under the acetate when producing the slide.

There are a number of ways of ensuring that letters and numbers are even: one is to use a stencil (which produces a professional-looking end result, but is slow); another is to use squared paper (with a suitable size of square) as a backing sheet and keep each letter or number within a single square; a third way is to write in the gap between two rulers which are kept parallel by sticking them together with sellotape leaving a gap between them equal to the desired letter height.

Use of viewgraphs

Obviously, to use any visual aid effectively, the speaker must have planned his or her talk in detail and have considered exactly when to show each slide.

Three further points are worth noting:

1. When a slide is projected, give the audience time to read it: either remain silent for a minute or two (which can feel like a very long time) or else read it through, slowly and clearly, to the audience. Never talk about something different; no-one can read one message and listen to another.

2. Do not leave a slide on view when you move on to talk about something else. If there is no other slide to be shown, switch the projector off (most projectors are cooled by an obtrusively noisy fan).

3. Finally, obviously—and most importantly—do not block the audience's view of the screen. A sensible precaution is to display your first viewgraph and then ask if everyone can see all of it. If they cannot, the problem can sometimes be solved by projecting the image higher, or by sitting rather than standing to deliver your talk.

Storage of OHP slides

Slides can be mounted in cardboard surrounds or simply left unmounted. The advantages of mounting slides in cardboard frames are:

—it helps preserve the slide in good condition

—overlays can be neatly hinged to the mount

—marks in the frames can be aligned with guides on the projector to ensure that the slide is centred and square-on in the projected image.

The disadvantages of mounting OHP slides are:

—it makes them heavy to carry around

—you may find that the mounted slides will not fit into a standard folder or briefcase.

35mm slides

The use of good 35mm slides, smoothly projected by remote control at the appropriate point, gives any presentation a thoroughly professional finish.

These slides will always be produced by a professional, though they will be designed by the speaker, and must therefore be planned well in advance. (By contrast, OHP slides can, if necessary, be drawn at home the night before a talk.) The same basic criteria of legibility and intelligibility which were listed in sections 1 and 2 on OHP slides apply here. In summary:

1. Slides must be clear and easy to read: no more than about six lines of text or numbers on a single slide.

2. They must be intelligible: each slide should contain only a single clear message.

3. If a story is to be built up by elaborating on an initial slide, a series of slides must be prepared, each one advancing the story a single step at a time.

35mm slides are loaded into a magazine (or carousel) before the talk in the order in which they will be used and are then projected onto the screen (or blank patch of wall) using a remote control button at the appropriate points during the talk. In general, lights must be dimmed to ensure that the slides are clearly visible.

By comparison with viewgraphs, 35mm slides have two considerable advantages:

—They look more professional.

—They are light and compact (several hundred 35mm slides will fit into a chocolate box).

They also have a number of disadvantages (in addition to the need to plan them well in advance):

—It is virtually impossible to alter the order in which they appear: this can be desirable if you have to adapt your presentation in response to unexpected questions from the audience, and it can be achieved relatively smoothly with viewgraphs.

—Similarly it is difficult to recall a slide from the middle of your talk if you wish to review major points at the end.

—It is not possible to use overlays or reveals with 35mm slides (though, of course, a similar effect can be achieved by using a series of slides, each one showing a little more of the story than the previous one).

—You cannot write on 35mm slides.

—You cannot tell by looking at the carousel of slides which slide will appear next. (By contrast, if viewgraphs are stored with a sheet of plain paper behind each, you can see the content of the next slide at a glance.)

Wall charts

Whereas viewgraphs and 35mm slides are normally used to illustrate a sequence of points throughout a talk, each slide appearing before the audience comparatively briefly, wall charts (or large charts supported on an easel) are more appropriate for a table or diagram which will be permanently on display throughout the talk and which will be referred to frequently. Such charts can therefore be used to display information like:

—a map of the geographical area from which data were collected and analysed, subdivided to show regions for which separate results were collected;

—a table showing a list of departments (or firms or areas) which have been studied, along with some basic data relevant to the presentation (for example, the number of staff in each department);

—an outline summary of the presentation showing the structure of the talk but omitting any results or details.

Three basic rules apply to the production and use of wall charts.

1. Make sure it is *intelligible*. This involves ensuring that the chart is clearly labelled and also explaining its contents to the audience at an early point in the presentation.

2. Make sure it is *legible*. This means that the chart should be produced using clear colours (not pink, orange, yellow or any pastel shade) and drawn with a thick pen or crayon. The amount of information on the chart must be limited to what can be read easily by the remotest member of the audience. (Clearly this means that more information can be included on a wall chart designed for use in a meeting of six than on one designed for a talk given to an audience of 60.)

3. Make sure it is *visible* without being obtrusive. The chart should be placed where each member of the audience has an uninterrupted view of it and where the speaker can point to it easily. Particularly when other visual aids are also used, wall charts or charts supported on easels should be placed where they will not dominate the presentation. If possible, such charts should be displayed at one side of the speaker's area rather than directly behind the speaker.

Appendix B: Further reading

Demonstration tables
EHRENBERG, A S C, *Data Reduction.* (chapters 1–4.) Wiley, 1975.

Reference tables
We have not found any general text on this subject. Departmental (or company) guidelines may be available in print: if so they should be considered carefully alongside the recommendations offered in Chapter 5.

Charts
SCHMID, C F, *Statistical Graphics: design principles and practices.* Wiley, 1983.

SCHMID, C F and SCHMID, S E, *Handbook of Graphic Presentation.* 2nd ed., Wiley, 1979.

ZELAZNY, Gene *Choosing and Using Charts.* New York, Video Arts, 1972.

Visual aids
American Society of Mechanical Engineers. *Illustration for Publication and Projection.* American National Standards Institution, 1979. (Y15. 1M–1979).

LAMB, Brydon, *Filmstrip and Slide Projectors in Teaching and Training*, Educational Foundation for Visual Aids, 1971.

POWELL, L S, *A Guide to the Overhead Projector*, British Association for Commercial and Industrial Education, 2nd ed., 1970.

Words
GOWERS, Sir E, *The Complete Plain Words*, revised by Sir Bruce Fraser, HMSO, 1973.

Plain English, a pamphlet published by the Cabinet Office (MPO), 1984.

FLETCHER, J, *How to Write a Report*, Institute of Personnel Management, 1983.

GUNNING, R, *The Technique of Clear Writing*, McGraw Hill, 1952.

A tape-slide course *Put it in Writing*, prepared by TFI with Albert Joseph.

Appendix C: Research evidence

This appendix is included in the belief that it is important to indicate the reasons for advocating the guidelines offered in the main text. Few of the guidelines are unsubstantiated by evidence that they work well and, in some cases, there is considerable evidence in their favour.

An excellent review article of research in this area was produced by Michael Macdonald Ross[1], and the interested reader is recommended first to that source.

A brief summary of the research evidence found in a literature review follows, along with references to the original sources.

Appendix D gives precise references to these sources and also to other articles which were consulted in preparing this book.

Summary of research evidence

1. For REFERENCE PURPOSES use TABLES (Washburne, 1927; Carter, 1947; Wainer et al, 1980):

   —items are easier to find by searching down a column than along a row (Wright and Fox, 1970)

   —break long columns into blocks of about five items (Tinker, 1960; Wright, 1968; Wright and Fox, 1970)

   —position columns close together (Wright and Fox, 1970).

2. Do NOT use TEXT ALONE to communicate quantitative information (Washburne, 1927; Feliciano et al, 1963).

3. If the reader is required to register SPECIFIC AMOUNTS use TABLES rather than charts (Washburne, 1927; Wainer et al, 1980).

4. In TABLES for DEMONSTRATION use HEAVILY ROUNDED numbers (Washburne, 1927; Ehrenberg, 1977 plus list of associated references).

5. Arrange TABLES for DEMONSTRATION so as to HIGHLIGHT PATTERNS (justification from the field of cognitive psychology, e.g. De Groot, 1966).

6. Except for reference tables, all tables and charts should be accompanied by a VERBAL SUMMARY (Feliciano et al, 1963; and 'depth of processing' argument from cognitive psychologists e.g. Craik and Tulving, 1975).

7. SIMPLE CHARTS are more effective than complex ones (Peterson and Schramm, 1955).

8. For comparison of RELATIVE MAGNITUDES of related measurements use BAR CHARTS—either vertical or horizontal bars—(Washburne, 1927; Feliciano et al, 1963) or ISOTYPES (Wrightstone, 1936). Isotype charts are likely to have more popular appeal than bar charts (Wrightstone, 1936).

8a. Grouped bar charts are better than segmented (component) bar charts if the composition of individual totals is to be shown (Culbertson and Powers, 1959). But see 9 below if change in composition is of greatest interest.

9. To illustrate the COMPOSITION of one or more totals use PIE CHARTS or PERCENTAGE COMPONENT BAR CHARTS (Culbertson and Powers, 1959; Peterson and Schramm, 1955) but

9a. Do NOT attempt to display differences in magnitudes of totals by using pies of different areas (Brinton, 1914; Flannery, 1971; Meihofer, 1973).

10. To display TIME TRENDS either LINE GRAPHS or VERTICAL BAR CHARTS may be used (Washburne, 1927; Schutz, 1961) but

10a. Surface (or layer) charts are best avoided (Culbertson and Powers, 1959).

11. Wherever possible lines and sections of pies or bars should be LABELLED DIRECTLY rather than by reference to a key (Culbertson and Powers, 1959).

12. Non-specialist audiences are likely to find graphics, generously supported by text, more appealing than tables (Feliciano, 1963).

13. People must be TRAINED if they are to understand the conventions used in constructing statistical charts (Vernon, 1953; Playfair, 1805).

---

[1] MACDONALD ROSS, M, How numbers are shown: A review of research on the presentation of quantitative data in texts. *Audio Visual Communications Review*, 1977, vol 25, no 4, pp. 359–407.

Appendix D: Bibliography

American Society of Mechanical Engineers. *Illustrations for Publication and Projection*. American National Standards Institute, 1979. (Y15. 1M–1979).

BEEBY, A W and TAYLOR, H P J. How well can we use graphs? *Communicator of Scientific and Technical Information*, 1973, vol 17, pp 7–11.

BRINTON, W C. *Graphic methods for presenting facts*. New York: The Engineering Magazine Company, 1915.

CARTER, L F. An experiment on the design of tables and graphs. *Journal of Applied Psychology*, 1947, vol 31 no 6, pp 640–650.

CRAIK, F I M and TULVING, E. Depth of processing and the retention of words in episodic memory. *Journal of Experimental Psychology: General*, 1975, vol 104, pp 268–294.

CROXTON, F E. Further studies in the graphic use of circles and bars Part 2: some additional data. *Journal of the American Statistical Association*, 1927, vol 22 no 1, pp 36–39.

CROXTON, F E and STEIN, H. Graphic comparison by bars, squares, circles and cubes. *Journal of the American Statistical Association*, 1932, vol 27 no 1, pp 54–60.

CROXTON, F E and STRYKER, R E. Bar charts versus circle diagrams. *Journal of the American Statistical Association*, 1927, vol 22, pp 473–82.

CULBERTSON, H M and POWERS, R D. A study of graph comprehension difficulties. *Audio Visual Communication Review*, 1959, vol 17, pp 97–110.

EELLS, W C. The relative merits of circles and bars for representing component parts. *Journal of the American Statistical Association*, 1926, vol 21 no 2, pp 119–132.

EHRENBERG, A S C. Graphs or tables? *The Statistician*, 1978, vol 27 no 2, pp 87–96.

EHRENBERG, A S C. Rudiments of numeracy (with discussion). *Journal of the Royal Statistical Society, Series A: General*, 1977, vol 140 para 3, pp 277–297.

EHRENBERG, A S C. 'What we can and can't get from graphs, and why.' (Paper based on invited talk at Statistical Meetings, Detroit, August 1981). 1982.

FEINBERG, S E. Graphical methods in statistics. *The American Statistician*, 1979, vol 33 no 4, pp 165–178.

FELICIANCO, G D, POWERS, R D and KEARL, B E. The presentation of statistical information. *Audio Visual Communication Review*, 1963, vol 11, pp 33–39.

FLANNERY, J J. The relative effectiveness of some common graduated point symbols in the representation of quantitative data. *Canadian Cartographer*, 1971, vol 8, pp 96–109.

FLETCHER, J. *How to Write a Report*. Institute of Personnel Management, 1983.

GOWERS, Sir E. *The Complete Plain Words*, revised by Sir Bruce Fraser. HMSO, 1973.

GREGG, V. *Human Memory*. Methuen, 1975.

DE GROOT, A D. Perception and memory versus thought: some old ideas and recent findings. *In* B Kleinmuntz (ed). *Problem Solving*. Wiley, 1966.

GROPPER, G L. Why IS a picture worth a thousand words? *Audio Visual Communication Review*, 1963, vol 11, pp 75–95.

HABER, R N. How we remember what we see. *Scientific American*, 1970, vol 222 part 2, pp 104–112.

HAMMERTON, M. How much is a large part? *Applied Ergonomics*. 1976, vol 7 no 1, pp 10–12.

HARTLEY, J and BURNHILL, P. Fifty guidelines for improving instructional text. *Programmed Learning and Education Technology*, 1977, vol 14 no 1, pp 65–73.

HYDE, J S and JENKINS, J J. Recall of words as a function of semantic, graphic and syntactic orienting tasks. *Journal of Verbal Learning and Verbal Behaviour*, 1973, vol 12 no 1, pp 147–180.

MACDONALD ROSS, M. How numbers are shown: a review of research on the presentation of quantitative data in texts. *Audio Visual Communication Review*, 1977, vol 25 no 4, pp 359–407.

MACDONALD ROSS, M. *Research in graphics communication: graphics in text, how numbers are shown*. 1977, Institute of Educational Technology, Monograph no 7, Open University.

MILLER, G A. The magical number seven, plus or minus two. *Psychological Review*, 1956, vol 63 no 2, pp 81–97.

MEIHOEFER, H J. The visual perception of the circle in thematic maps: experimental results. *Canadian Cartographer*, 1973, vol 10, pp 63–84.

NEURATH, M. Isotype. *Instructional Science*, 1974, vol 3 no 2, pp 127–150.

NEURATH, O. *International Picture Language*, Kegan Paul, Trench and Trubner, 1936.

NEURATH, O. *Modern man in the making*. Knopf, New York, 1939.

NICKERSON, R S. Short-term memory for complex meaningful visual configurations: a demonstration of capacity. *Canadian Journal of Psychology*, 1965, vol 19 no 2, pp 155–160.

PAIVIO, A and CSAPO, K. Concrete image and verbal memory codes. *Journal of Exptl. Psychology*, 1969, vol 80 no 2, pp 279–285.

PEARSON, E S. Some aspects of the geometry of statistics. *Journal of the Royal Statistical Society, Series A: General.* 1956a, vol 119 no 2, pp 125–149.

PETERSON, L V and SCHRAMM, W. How accurately are different kinds of graphs read? *Audio Visual Communication Review*, 1955, vol 2 no 2, pp 178–189.

PLAYFAIR, W. *The Commercial and Political Atlas*, 3rd ed. London: J Wallis, 1801.

SCHUTZ, H G. An evaluation of formats for graphic trend displays (Parts A and B). *Human* Factors, 1961, vol 3, pp 99–119.

SCHMID, C F. *Statistical Graphics: design principles and practices.* Wiley, 1983.

SCHMID, C F and SCHMID, S E. *Handbook of Graphic Presentation*, 2nd ed. Wiley, 1979.

SHEPARD, R N. Recognition memory for words, sentences and pictures, *Journal of Verbal Learning and Verbal Behaviour*, 1967, vol 6 no 1, pp 156–163.

SIMON, H A. *Models of Thought.* Yale University Press, 1979.

SIMON, H A and GREGG, L W. One trial and incremental learning. Chapter 3.3 of *Models of Thought* by H A Simon. Yale University Press, 1967.

TINKER, M A. Legibility of mathematical tables. *Journal of Applied Psychology*, 1960, vol 44 no 2, pp 83–87.

TUKEY, J W. *Exploratory Data Analysis.* Addison-Wesley, 1977.

TULVING, E. Subjective organisation and effects of repetition in multi-trial free-recall learning. *Journal of Verbal Learning and Verbal Behaviour*, 1966, vol 5 no 2, pp 193–197.

STANDING, L, CONEZIO, J and HABER, R N. Perception and memory for pictures: single trial learning of 2560 stimuli. *Psychonomic Science*, 1970, vol 19, pp 73–74.

VERNON, M D. Learning from graphical material. *British Journal of Psychology*, 1946, vol 36, pp 145–158.

VERNON, M D. The use and value of graphical material in presenting quantitative data. *Occupational Psychology*, 1952, vol 26 no 1, pp 22–34 and no 2, pp 96–100.

VERNON, M D. Presenting information in diagrams: *Audio Visual Communication Review*, 1953, vol 1, pp 147–158.

WAINER, H, LONO, M and GROVES, C. *On the display of data: some empirical findings.* 1980. Washington DC: Bureau of Social Science Research. Draft report.

WASHBURNE, J N. An experimental study on various graphic, tabular and textual methods of presenting quantitative material. *Journal of Educational Psychology*, 1927, vol 18 no 6, pp 361–376 and no 7, pp 465–476.

WAUGH, N C. Retrieval time in short-term memory. *British Journal of Psychology*, 1970, vol 61 no 1, pp 1–12.

WAUGH, N C and NORMAN, D A. Primary memory. *Psychological Review*, 1965, vol 72 no 2, pp 89–104.

WEINTRAUB, S. What research says to the reading teacher (graphs, charts and diagrams). *Reading Teacher*, 1967, vol 20, pp 345–349.

WRIGHT, P. Using tabulated information. *Ergonomics*, 1968, vol 11 no 4, pp 331–343.

WRIGHT, P and FOX, K. Presenting information in tables. *Applied Ergonomics*, 1970, vol 1 no 4, pp 234–242.

WRIGHTSTONE, J W. Conventional versus pictorial graphics. *Progressive Education*, 1936, vol 13 no 6, pp 460–462.

ZELAZNY, G. *Choosing and Using Charts.* London. Video Arts, 1972.

Printed in UK for HMSO
Dd 737491 C50 7/86

SOCIAL TRENDS

For sixteen years Social Trends has provided a valuable insight into life in Britain and its changes

This new edition of Social Trends updates its broad description of British society. The material in it is arranged in chapters corresponding closely to the administrative functions of Government. The focus in each chapter is on current policy concerns. Latest available data are included wherever possible.

Social Trends not only is necessary for people involved in social policy and social work both in government and outside government but is also an invaluable guide for market researchers, journalists, teachers, advertisers, businessmen — anyone, in fact, who has a concern for British society.

Social Trends 16

For the fifth year running price held at £19.95

ISBN 0 11 620151 7

Central Statistical Office publications are published by Her Majesty's Stationery Office.
They are obtainable from Government bookshops and through good booksellers.